3rd International Symposium on Atomic Technology 2009

AA009051

Journal of Physics: Conference Series Volume 191

Tokyo, Japan
5-6 March 2009

ISBN: 978-1-61738-285-7

Printed from e-media with permission by:

Curran Associates, Inc.
57 Morehouse Lane
Red Hook, NY 12571

Some format issues inherent in the e-media version may also appear in this print version.

Copyright© (2009) by the Institute of Physics
All rights reserved.

Printed by Curran Associates, Inc. (2010)

For permission requests, please contact the Institute of Physics
at the address below.

Institute of Physics
Dirac House, Temple Back
Bristol BS1 6BE UK

Phone: 44 1 17 929 7481
Fax: 44 1 17 920 0979

techtracking@iop.org

Additional copies of this publication are available from:

Curran Associates, Inc.
57 Morehouse Lane
Red Hook, NY 12571 USA
Phone: 845-758-0400
Fax: 845-758-2634
Email: curran@proceedings.com
Web: www.proceedings.com

TABLE OF CONTENTS

ATOMIC AND POLYSCALE TECHNOLOGIES FOR BIOMEDICAL APPLICATIONS

ATOMIC AND POLYSCALE FABRICATION FOR ADVANCED MATERIALS AND DEVICES

QUANTUM AND MOLECULAR ENGINEERING FOR ADVANCED TECHNOLOGIES

The International Symposium on Atomic Technology (ISAT) is held every year. The Third Symposium (ISAT-3) was held on 5–6 March 2009 at the Tokyo International Exchange Center, Tokyo, Japan jointed with the Third Polyscale Technology Workshop (PTW-3).

The ISAT-3 symposium was intended to offer a forum for the discussion of the latest progress in atomic technologies, which was successively held after ISAT-1 at Tsukuba and ISAT-2 at Awaji in 2007. The symposium was attended by 136 participants. There were 12 invited and 4 oral presentations. The number of poster presentations was 101. From all the contributions, 32 papers selected through review process are contained in this volume.

The 'Atomic Technology Project' and the 'Polyscale Technology Project' were started in 2006 as the joint project of three institutions; (1) Center for Atomic and Molecular Technologies, Graduate School of Engineering, Osaka University (CAMT), (2) Tsukuba Research Center for Interdisciplinary Materials Science, Graduate School of Pure and Applied Sciences, University of Tsukuba (TIMS) and (3) Polyscale Technology Research Center, Research Institute for Science and Technology, Tokyo University of Science (PTRC), each of which were independently pursuing atomic and polyscale technologies. The project is funded by the Ministry of Education, Culture, Sports, Science and Technology of Japan.

The goal of these projects is to contribute to the development of atomic and polyscale science and technologies. In this symposium, four research fields were focused on: Biomedical Applications, Fabrication for Advanced Materials and Devices, Magnetic Applications, and Quantum and Molecular Engineering for Advanced Technologies.

Atsuo Yasumori
Conference Chair
Polyscale Technology Research Center, Research Institute for Science and Technology, Tokyo University of Science, 2641 Yamazaki, Noda, Chiba 278-8510, Japan.

All papers published in this volume of *Journal of Physics: Conference Series* have been peer reviewed through processes administered by the proceedings Editors. Reviews were conducted by expert referees to the professional and scientific standards expected of a proceedings journal published by IOP Publishing.

Third International Symposium on Atomic Technology IOP Publishing
Journal of Physics: Conference Series **191** (2009) 012001 doi:10.1088/1742-6596/191/1/012001

Selective Tumor Imaging by a Novel Tumor Specific Aralin-Infrared-to-Visible Phosphor Conjugate

Y Kawasaki[1,2], Y Gotoh[1,2], K Tokuzen[2,3], M Kamimura, [4,5], T Komeno[1], M Tomatsu [6], R Todoroki[1,2], Y Nagasaki[2,4,5], K Soga[2,3] and F Tashiro[1,2]

[1] Department of Biological Science and Technology, Tokyo University of Science,

2641 Yamazaki, Noda, 278-8510 Chiba, Japan

[2] Polyscale Technology Research Center,

2641 Yamazaki, Noda, 278-8510 Chiba, Japan

[3] Department of Materials Science and Technology, Tokyo University of Science,

2641 Yamazaki, Noda, 278-8510 Chiba, Japan

[4] Graduate School of Pure and Applied Science, University of Tsukuba,

1-1-1 Tennoudai, Tsukuba, 305-8573, Ibaraki, Japan

[5] Tsukuba Research Center for Interdisciplinary Materials Science (TIMS),

1-1-1 Tennoudai, Tsukuba, 305-8573, Ibaraki, Japan

[6] Akita Research Institute for Food and Brewing (ARIF),

4-26 Azasanuki Arayamachi, Akita, 010-1623 Akita, Japan

E-mail: ftashir@rs.noda.tus.ac.jp

Abstract. Aralin is a novel cytotoxic protein from *Aralia elata* and selectively induces apoptosis in transformed cells as compared to normal cells (1). Aralin is a lectin specific for sugar chain such as galactose and possesses RNA *N*-glycosidase activity. In this study, antitumor potency of aralin was analyzed using the poly(ethyleneglycol) (PEG)/streptavidin co-immobilized infrared-to-visible upconversion phosphors, Y_2O_3 nanoparticles (2). Cy3-conjugated aralin could clearly detect the surface of SV40-transformed VA13 and human cervical carcinoma HeLa cells, but to a lesser extent on the normal human fibroblast WI-38 cells. Conjugation of aralin with PEGylated Y_2O_3 nanophosphor was carried out via biotin-avidin binding. The Y_2O_3-conjugated aralin also clearly visualize by a fluorescence microscope measurements equipped with near-infrared excitation source scanning in HeLa cells. It is also important to note that no remarkable damage to the cells was observed during these observations. Thus, these data imply that the Y_2O_3-conjugated aralin would potentially be useful material for tumor detection *in vivo*.

Journal of Physics: Conference Series **191** (2009) 012001 doi:10.1088/1742-6596/191/1/012001

1. Introduction

Visualization of the tumor cells is one of the most important techniques to find out cancer patient in early stage of the disease (3). We have been reported that the novel type II ribosome-inactivating protein aralin from *Aralia elata*, is an anti-tumor cytotoxic protein (1). Aralin is composed of A- and B-chains, whose N-terminal sequences show homologies with nigrin A-chain and ricin B-chain, respectively, and highly toxic to tumor cell lines compared with normal cell lines (1). Since aralin possesses a lectin-like specificity for galactose, lactose and fucose, it seems that aralin recognizes some types of tumor cell surfaces, in which sugar-added receptors are expressed to facilitate the delivery of toxic A-chain by endocytosis (4). After the entry into the cytosol, the A-chain inactivates 60S ribosomal subunits by catalyzing specific RNA-*N*-glycosidase activity and inhibited protein synthesis (4). Therefore, aralin may prove to be a suitable immunotoxin for tumor therapy and detection.

In order to visualize such specific toxicity against cells and molecular events in the cytosol, fluorescence bioimaging system is one of the most powerful tools. However, ultraviolet light causes toxic effect to living organism including cells. If near infrared light can be utilized for this system, the bioimaging technique will be improved. This paper communicates biolabeling of aralin by novel PEGylated nanophosphors, which can be excited by near infrared laser. The conjugate thus obtained was applied for an *in vitro* experiment to confirm antitumor potency.

2. Experimental Section

2.1. Cytotoxicity assay

Normal and cancer cell lines were seeded at 1 x 10^4 cells/well in 24-well plate and cultured in Eagle's minimal essential medium (MEM) or Dulbecco's modified Eagle's medium (DMEM) supplemented with 10% FCS for 24 h. Various concentration of aralin were added to cells for 36 h and cytotoxity of aralin was examined by trypan blue dye exclusion assay.

2.2. Preparation of Cy3-conjugated aralin and detection of cell surface receptor

The N-terminal amino groups of aralin were labeled with Cy3 (GE healthcare, Piscataway, NJ, USA) according to the manufacture's procedures. Free Cy3 was removed by Sephadex G-50 gel filtration. Cells attached to Lab-Tek II chamber slides (Nalge Nunc International, Naperville, USA) and suspended in MEM with 10% FCS were incubated with 3.2 x 10^{-9} M Cy3-aralin at 4°C for 15 min and the cells were examined under a fluorescence microscope (Axioplan 2, Carl Zeiss, Oberkochen, Germany).

2.3. Preparation of PEGylated nanophosphor conjugated aralin and *in vitro* experiments

The N-terminal amino groups of aralin were labeled with biotin labeling kit-NH_2 (Dojindo, Kumamoto, Japan) according to the manufacture's procedures. The PEG/streptavidin modified infrared-to-visible upconversion phosphors, Y_2O_3 nanoparticles were prepared as previously described (4). Two hundred ng biotin-conjugated aralin and 1 μg PEG/streptavidin-Y_2O_3 were reacted at 4°C for 30 min. VA-13 cells were suspended with HEPES buffer (20 mM Hepes-KOH, pH7.5, 150 mM NaCl and 3% BSA) and incubated with biotin-aralin and PEG/streptavidin-Y_2O_3 at 4°C for 2 h. and examined under a fluorescence microscope equipped with near-infrared excitation source (TCLDM9, Thorlabs, Newton, NJ, USA).

3. Results and Discussion

3.1. Cytotoxic activity of aralin

To investigate the effects of aralin on the normal human fibroblasts WI-38 and cancer cells such as tumorigenic simian virus 40 large T antigen-transformed WI-38 cells (VA-13) and human cervical carcinoma HeLa cells, the cell viability was assayed by trypan blue dye exclusion procedure. Aralin exerted a potent cytotoxicity against human cancer cells (VA-13, HeLa) compared with WI-38 cells (Figure 1). The IC_{50} vales of aralin against WI-38, VA-13 and HeLa cells were 42.3, 4.0 and 1.6 ng/ml, respectively.

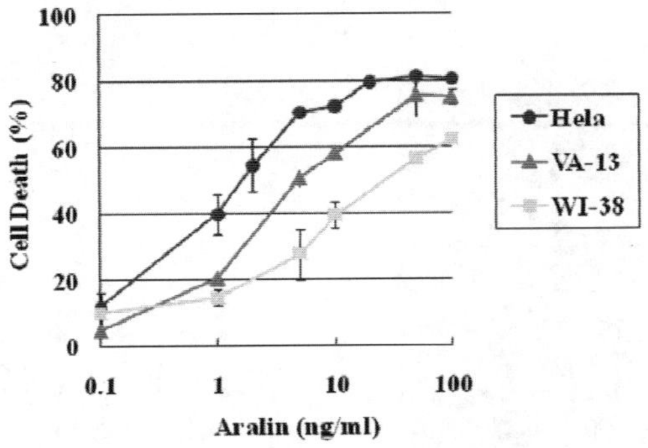

Figure 1. Cytotoxic activity of aralin in normal and various human tumor cells.
Normal and cancer cell lines were seeded at 1 x 10^4 cells/well in 24 well plate and cultured in MEM or DMEM supplemented with 10% FCS for 24 h. Various concentration of aralin were added to cells for 36 h and cytotoxity of aralin was examined using trypan blue dye exclusion assay.

3.2. Aralin binding with cell surface

The antitumor potency of arlain was then investigated using Cy3-conjugated aralin. When WI-38, VA-13 and HeLa cells were incubated with 3.2 x 10^{-9} M Cy3-aralin at 4°C for 15 min, the cell surface of VA-13 and HeLa cells were significantly labeled with Cy3-aralin compared with WI-38 cells, but not by Cy3-BSA (Figure 2). Fluorescence of Cy3-aralin was hardly detected in WI-38 cells under the same conditions, but slightly detected for further incubation at 4°C (data not shown). These results suggest that the expression level of the aralin receptor on the cell surface of WI-38 cells and/or its binding affinity are lower than those of VA-13 and HeLa cells.

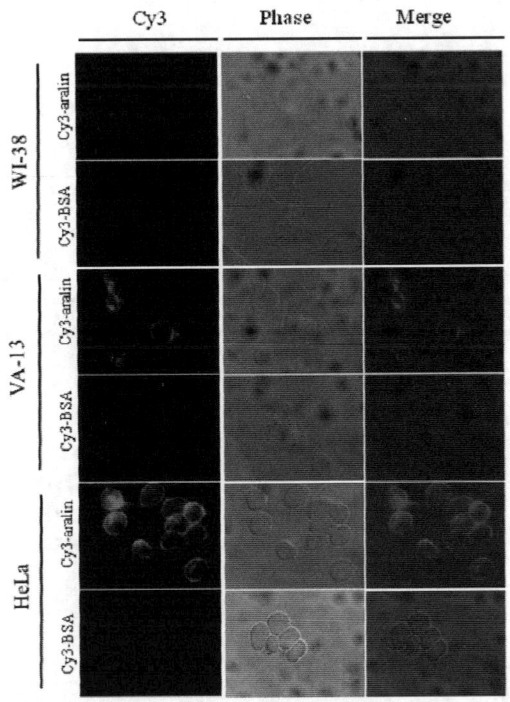

Figure 2. Detection of aralin cell surface receptor.
WI-38, VA-13 and HeLa cells were incubated with 3.2 x 10^{-9} M Cy3-conjugated aralin or BSA at 4°C for 15 min. The cells were observed under a fluorescence microscope.

3.3. Selective tumor imaging using PEGylated nanophosphor conjugated araline

In order to confirm near-infrared excitation system, PEGylated Y_2O_3 nanophosphor-conjugated aralin was investigated for the potent tumor imaging. VA-13 cells were incubated with 3.2 x 10^{-8} M PEGylated nanophosphor conjugaeted-aralin or BSA at 4°C for 2 h and examined under a fluorescence microscope equipped with near-infrared excitation source. As shown Figure 3, Y_2O_3 excitation was detected with PEGylated nanophosphor conjugaeted-aralin treated on cell surface, but not by PEGylated nanophosphor conjugaeted-BSA. Therefore, we achieved detection of tumor cells using anti-tumor protein and near infrared excitation phosphors Y_2O_3. It should be noted that the conjugate was excited by 980 nm laser (200 mW), thus no cell damage was observed in this study.

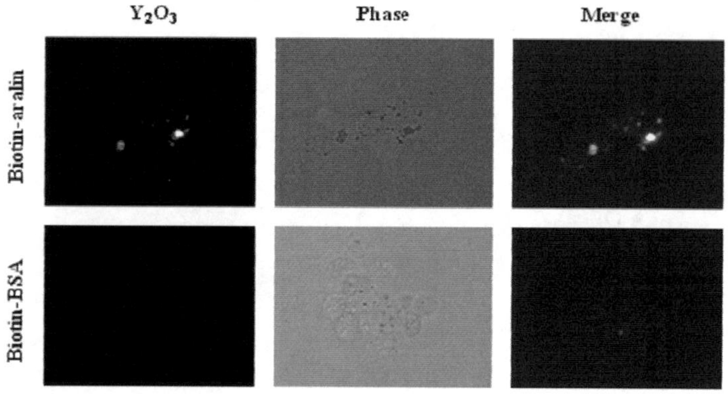

Figure 3. Selective tumor imaging using biotin-aralin and PEG/streptavdine-Y_2O_3
VA-13 cells were incubated with 3.2 x 10^{-8} M biotin-aralin or BSA at 4°C for 120 min. VA13 cells were applied on slides glasses and examined under a fluorescence microscope equipped with a 980-nm laser diode for the near-infrared excitation source.

4. Conclusion
These results indicate that the selective tumor cells detection systems using PEGylated nanophosphor conjugaeted-aralin would be expected as useful biolabeling materials for tumor detection *in vivo*.

5. Acknowledgemnts
This work was partly supported by "Academic Frontier" Project for Private Universities: Matching Fund Subsidy from MEXT (Ministry of Education, Cultures, Sports, Science and Technology), 2006-2010. This work was also supported by KAKENHI(17590098): Grant-in-Aid for Scientific Research (C), 2005-2006.

References
[1] Tomatsu M., Kameyama M.O., Shibamoto N. 2003 *Cancer letter* **199** 19
[2] Kamimura M., *et. al.* 2008 *Langmuir* **24** 8864
[3] Hoffman R.M. 2005 *Nat. rev. cancer* **5** 796
[4] Tomatsu M., *et. al.* 2004 *Biol. Chem.* **385** 819

Journal of Physics: Conference Series **191** (2009) 012002 doi:10.1088/1742-6596/191/1/012002

Bioimaging of M1 Cells Using Ceramic Nanophosphors: Synthesis and Toxicity Assay of Y_2O_3 Nanoparticles

N Venkatachalam[1], Y Okumura[2], K Soga[1, 2], R Fukuda[3] and T Tsuji[1, 3]

[1]Polyscale Technology Research Center, Tokyo University of Science

[2]Department of Materials Science and Technology, Tokyo University of Science

[3]Department of Biological Science and Technology, Tokyo University of Science

2641 Yamazaki, Noda, Chiba 278-8510, JAPAN

E-mail: venkat@sogalabo.jp

Abstract. Er^{3+} doped Y_2O_3 nanoparticles were synthesized by enzymatic and polymer-assisted homogeneous co-precipitation methods. Resultant particle size is about 30-40 nm with narrow size distribution whereas the particle size is smaller than those acquired by conventional homogeneous and alkali precipitation methods. The particles shows bright green (550 nm) and red (660 nm) upconversion (UC) as well as near infrared (NIR) fluorescence (1550 nm) under 980 nm excitation. Bioimaging of M1 cells using the nanoparticles were successfully attempted. It is observed that 0.5 mg/ml of nanoparticles is the nominal concentration for bioimaging of M1 cells under the physiological conditions. The cellular uptake of nanoparticles is evidenced from bright field, UC and NIR fluorescence images of live M1 cells. Our studies suggest that lower concentration of nanoparticles is sufficient for imaging when the particles are taken in the M1 cells and also the concentration can keep the cells alive. Further it was demonstrated that under the physiological conditions, Y_2O_3 nanoparticles emit UC and NIR fluorescence in M1 cells even after the surface modification with PEG-*b*-PAAc polymer. Moreover, surface modified nanoparticles shows lower toxic effect in M1 cells while compare to bare nanoparticles.

1. Introduction

Rare-earth (RE) doped inorganic nanostructure materials have attracted much attention in recent years because of their unique physical and chemical properties. Among the ceramic nanophosphors, Y_2O_3 is a promising candidate owing to its photo durability and its phonon energy that is low enough to reveal upconversion (UC) and near infrared (NIR) emission when trivalent RE ion doped in it [1,2]. Recently UC bioimaging using RE doped Y_2O_3 nanophosphors has been reported as a promising alternative candidate for fluorescent dyes in overcoming the color fading, photo toxicity, auto-fluorescence and light scattering problems that are encountered during fluorescence bioimaging of living cells with the use of ultraviolet or visible excitation lights [3]. The UC is a phenomenon that converts near infrared light into visible light by a multistep excitation process in discreet energy levels of trivalent RE ions in ceramic hosts. An important merit of the UC bioimaging is the usage of a NIR excitation light, which does not cause all of the above problems, such as color fading, photo toxicity etc. In order to apply the UC process to intra cellular bioimaging, the desirable size of the ceramic nanophosphors should be less than 100 nm and the size distribution should be narrow. To meet the requirements we have to control the three important synthesis conditions such as supersaturation, faster nucleation and slower diffusion of the starting materials as well as precipitated precursor particles during the homogeneous precipitation process. Because in a homogeneous method, water at 80 ^{0}C is very active and the diffusion of starting materials with respect to particle growth in such a water medium is fast. Instead of supplying activation energy through heating, one can use a biological catalyst such as enzymes in order to overcome such problems. Lower temperature (25 ^{0}C) of the reaction medium not only enhances the slower diffusion and growth of the nanoparticles but also avoids damage to the enzyme in an aqueous medium during the synthesis of precursor materials [4]. In another point of view, it is noteworthy that addition of capping agents such as polymers in an aqueous medium along with starting materials may enhance the viscosity and also creates some ionic interactions with metal ions which suppress the diffusion of starting materials as well as precipitated precursors during the homogeneous precipitation process [5]. Diffusion suppression ultimately leads to supersaturation of starting materials which favours the formation of large number of nucleation sites. In order to demonstrate the influence of polymer during the homogeneous co-precipitation of precursor particles [$Y(OH)CO_3^{2-}$], we have investigated the effect of addition of poly cationic [polyallylamine (PAA)] and poly anionic [polyacrylic acid (PAAc)] polymers. Our experimental results suggest that the presence of PAAc during the homogeneous co-precipitation of precursor particles effectively controls the particle size than the presence of PAA.

Bioimaging of M1 cells (macrophages) using Er^{3+} doped Y_2O_3 nanoparticles by UC as well as NIR fluorescence were observed under 980-nm excitation. UC bioimaging of M1 cells under NIR fluorescence is not reported to date. M1 cells are key players in the immune response to foreign invaders such as infectious microorganisms when it enters in our body. Likewise, they help to destroy bacteria, protozoa, and tumour cells [6-8]. Thus, we demonstrated the experimental work in M1 cells, which is commercially available cell line of mouse macrophage. The influence of various experimental parameters such as concentration of Y_2O_3 nanoparticles with respect to the survivability of macrophages (toxicity), incubation period (time required to uptake the nanoparticles by M1 cells), particle size effect, toxicity of bare and surface modified nanoparticles were investigated in order to improve the imaging process. Therefore, the present study demonstrates the synthesis, characterization of Er^{3+} doped Y_2O_3 nanoparticles as well as its UC and NIR fluorescence in M1 cells under 980-nm excitation along with the estimation of the cell toxicity of the nanoparticles in the cells.

2. Experimental

2.1. Preparation of Er^{3+} doped Y_2O_3 nanoparticles

The Er^{3+} doped Y_2O_3 nanoparticles were prepared by calcination of precursors at 1100 ^{0}C for 1 h. The precursors [$Y(OH)CO_3^{2-}$] were precipitated with enzymatic and polymer assisted homogeneous co-precipitation methods. A mixture of urea (0.4 mol), nitrates of yttrium and erbium were dissolved into distilled water (300 ml) as a starting solution. The amount of Er^{3+} source corresponds to the final composition of the particles to be $Y_{0.93}$ $Er_{0.07}$ $O_{1.5}$. For enzymatic decomposition method, urease enzyme (30 mg) was added to the starting solution and for homogeneous co-precipitation method, PAAc or PAA (0.1 mmol) was dissolved into the starting solution. By heating (for homogeneous precipitation) the solution at 80 ^{o}C for 2 h, the urea decomposed into CO_3^{2-} and NH_4^+ as precipitants for precursor precipitation. For enzymatic decomposition of urea, the solution was kept for stirring under room temperature (25 ^{0}C) for 2 h. For conventional alkali precipitation method, aqueous NH_4OH (12 ml from 1.0 mol/L) was added to the starting solution and kept for stirring at 80 ^{o}C for 2 h. The precipitated precursors were washed three times with distilled water centrifugally and subsequently washed one time with alcohol and then dried at 100 ^{0}C for 12 h. The dried precursor materials were calcined in air atmosphere at 1100 ^{0}C for 1 h.. Precipitated precursor materials as well as calcined Y_2O_3 nanoparticles were characterized by XRD, FT-IR, TG-DTA, DLS, FE-SEM and TEM observations.

Journal of Physics: Conference Series **191** (2009) 012002 doi:10.1088/1742-6596/191/1/012002

2.2. UC and NIR emission spectrum measurements

UC and NIR emission spectrum of the samples was measured using a spctroflurophotometer equipped with a diode laser at 980-nm with 260 mW power as an excitation source.

2.3. UC and NIR fluorescence bioimaging

UC and NIR fluorescence images of the samples were observed with biological microscope equipped with a near infrared excitation (980-nm) as described above and Si and InGaAs-CCD cameras for the UC and NIR fluorescence imaging respectively.

2.4. Bioimaging of M1 cells

Commercially available M1 cell lines of mouse were used for the experiments. For the cellular uptake of Y_2O_3 nanoparticles by M1 cells, various concentrations (0.1-3 mg/ml) of nanoparticles with an average size of 200 nm was mixed with culture medium (mixture of D-MEM cell culture medium and 5% horse serum). After that, matured M1 cells are added to the culture medium and incubated for 12, 24 and 48 h at 37 ^{0}C for the cellular uptake of nanoparticles. UC and NIR fluorescence of the M1 cells were observed with above described biological fluorescence microscope.

3. Results and Discussion

3.1. Physico-chemical characterization of nanoparticles

Figure 1 shows FE-SEM and TEM images of precursors as well as calcined Y_2O_3 nano particles obtained by various preparation methods. Among the methods enzymatic and polymer assisted co-precipitation methods by using PAAc yield amorphous precursors with a size of 20-30 nm within a narrow size distribution. Figure 2 shows the particle size distribution of precursor particles and calcined Er^{3+} doped Y_2O_3 nanoparticles. We observed that even after calcination process there is no greater change in size and morphology of the nanoparticles which were evident from DLS, FE-SEM and TEM observations when the enzymatic and polymer-assisted methods are used, while obvious growth of the particles was observed when used with homogeneous and alkali precipitation methods. The authors have reported that there is not an obvious change in the morphology and size when the precursors were calcined at 900 ^{o}C for 1 h [9]. The calcination at 1100 ^{o}C for 1 h, causes the crystal growth though it is proved that the growth can be avoided by the use of enzymatic and polymer-assisted methods. It was observed that the physico-chemical characterization of precursor particles results were in good agreement with our previous report [9]. On the basis of the characterization studies, we concluded that the precipitated precursors are in the form of amorphous yttrium hydroxyl carbonate [$Y(OH)CO_3^{2-}$]. Moreover, precursor samples co-precipitated with PAAc shows the characteristic vibrations for the presence of CH_2 and COO^- molecules even after washing processes which were evident from FT-IR and TG-DTA analysis. This may be due to the fact that PAAc is an ionomer with many negatively charged carboxylate anion (COO^-) groups; it has the significant effect to attract Y^{3+} and Er^{3+} cations in an aqueous medium. Thus, we believe that the diffusion of cations is suppressed significantly by the presence of PAAc than the presence of PAA. Because PAA is a polycationic polymer; it makes repulsion with Y^{3+} and Er^{3+} cations in an aqueous medium. It was demonstrated that calcination process eventually removes the organic residues, carbonate and hydroxide species from the Y_2O_3 nanoparticles which were evident from FT-IR, TG-DTA and SEM-EDAX observations.

Figure 1 FE-SEM images of $Y(OH)CO_3^{2-}$ particles precipitated by (a) enzymatic, (b) homogeneous precipitation, (c) polymer- assisted homogeneous co-precipitation and (d) alkali precipitation method. FE-SEM (c1) and TEM (c2) images of Y_2O_3 nanoparticles were observed after the calcination of $Y(OH)CO_3^{2}$ particles precipitated by the method (c) are also shown .

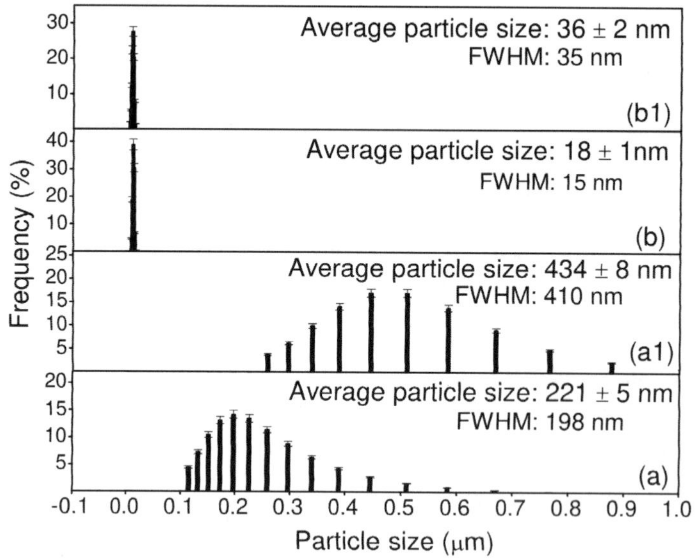

Figure 2 Particle size distributions of as-synthesized and calcined $Y(OH)CO_3^{2-}$ particles obtained by the DLS measurement. (a) and (a1) are as-synthesized and calcined Y_2O_3 nanoparticles by conventional homogeneous precipitation method respectively. (b) and (b1) are as-synthesized and calcined Y_2O_3 nanoparticles by polymer- assisted homogeneous co- precipitation method respectively.

3.2. UC and NIR emission of Er^{3+} doped Y_2O_3 nanoparticles

Figures 3 and 4 show the UC and NIR emission spectra as well as fluorescence images of Er^{3+}doped Y_2O_3 nanoparticles (sample b1). It was confirmed that under physiological conditions Er^{3+} ions emits UC and NIR fluorescence under 980-nm excitation.

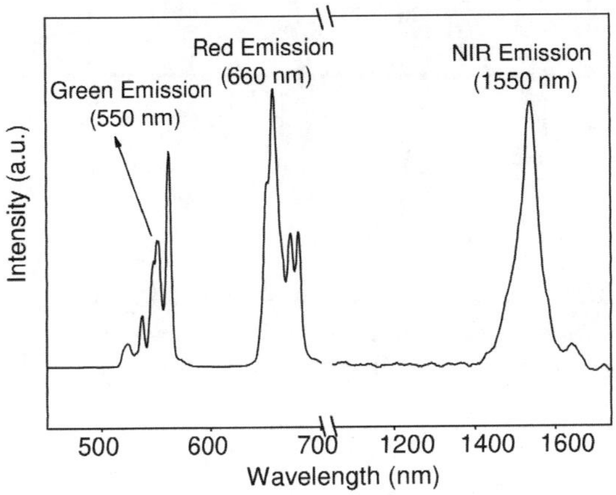

Figure 3 UC emission spectra of sample (b1) under 980- nm excitation.

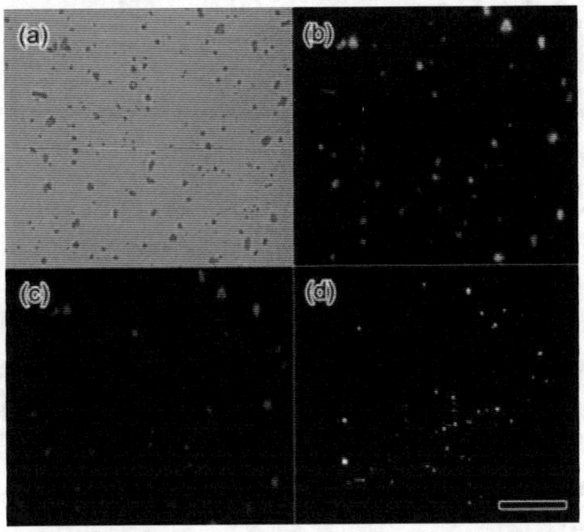

Figure 4 Bright field (a), green (b) and red (c) upconversion and NIR (d) fluorescence images of the sample b1 in physiological buffer under 980-nm excitation in biological microscopic view (scale bar: 25 μm).

3.3. UC and NIR imaging and toxicity studies in M1 cells

The efficiency of UC bioimaging of M1 cells depends upon the various factors including the viability of the M1 cells after up-taking the Y_2O_3 nanoparticles. Hence it is essential to optimise the Y_2O_3 concentration to the live M1 cells under physiological conditions in order to achieve the efficient bioimaging. It is observed that the amount of Y_2O_3 nanoparticles taken by M1 cells increases with increase of Y_2O_3 nanoparticles concentration up to 0.5 mg/ml (more than 80 % cells are alive when compare to the control sample) and above the concentration the cell viability decreased (Figure 5). Furthermore we found that longer incubation period (upto 48 h) does not affect the viability of the M1 cells under the above optimised concentration (Figure 5). Besides, we investigated the size effect under the same experimental condition (0.5 mg/ml of Y_2O_3; 24 h incubation). We observed that the cell death rate marginally increased when the particle size decreases (Figure 6). This may be due to the fact that smaller size Y_2O_3 nanoparticles start to dissolve gradually when it retains longer time within the M1 cells. We suspect that, M1 cells food path is under acidic nature so that Y_2O_3 nanoparticles starts to dissolve and may increase the toxicity in M1 cells. This may also retards the fluorescence of nanoparticles when we want to observe the M1 cells after the longer incubation period like above 48 h. In order to increase the chemical durability of Y_2O_3 nanoparticles we made the surface modification with PEG-*b*-PAAc [poly (ethylene glycol)-*b*-poly (acrylic acid)]. Recently our research group members are successfully reported the surface modification of Y_2O_3 nanoparticles using PEG-*b*-PAAc [10-12]. It is also reported that another PEG modification on the Y_2O_3 nanoparticles by using a silane coupling agent is another effective surface modification method [12, 13]. We observed that the chemical durability of Y_2O_3 particles was improved significantly after the surface modification with PEG-*b*-PAAc. Moreover, after the surface modification, agglomeration of the nanoparticles in M1 cells was significantly reduced under the physiological conditions when compare to bare particles. Hence, surface modified Y_2O_3 nanoparticles with PEG-*b*-PAAc shows lower toxic effect in M1 cells than the bare Y_2O_3 nanoparticles (Figure 6).

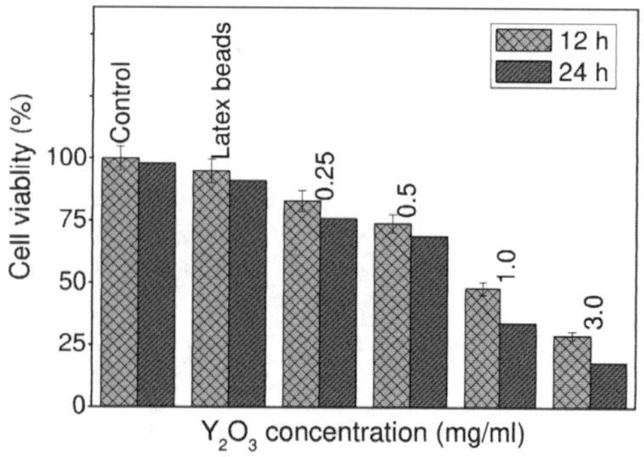

Figure 5 Cell toxicity of Y_2O_3 nanoparticle with various concentrations. The number in the view graph denotes the Y_2O_3 nanoparticles in mg/ml. The result of the fluorescent latex beads is also shown for a comparison.

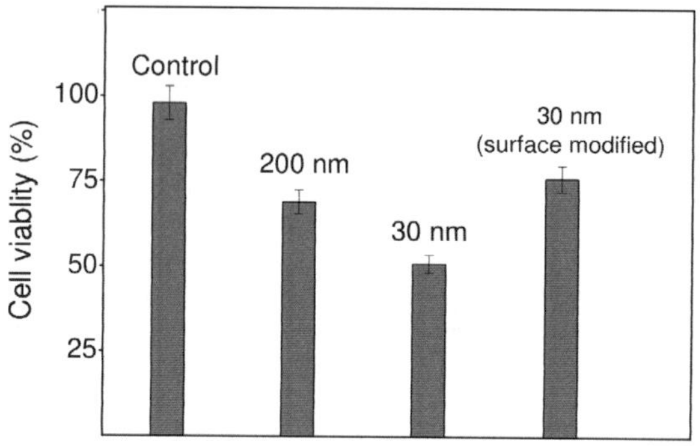

Figure 6 Cell toxicity of bare and surface modified Y_2O_3 nanoparticle with various sizes. The concentrations of the nanoparticles are fixed to be 0.5 mg/ml.

The cellular uptake of nanoparticles is evidenced from the bright field (Figures 7a & 7b), UC and NIR fluorescence (Figure 7c and 7d) images of the live M1 cells incubated for 24 h. To compare the efficiency of Y_2O_3 nanoparticles with the conventional fluorescence agent, some of the experiments were done with commercially available latex beads. The cellular uptake of latex beads amount is higher when compare to Y_2O_3 nanoparticles under the identical experimental conditions. However, the emission intensity of the latex beads was diminished within several tens of minutes whereas the Y_2O_3 nanophosphors maintain a bright image for more than 12 h. Our studies suggest that even lower concentration of Y_2O_3 nanoparticles (0.5 mg/ml) is sufficient for the UC imaging when the particles are taken in the cells, and also the concentration can keep the M1 cells alive. In order to improve the efficiency of the UC as well as NIR bioimaging under 980-nm excitation further experimental investigations are under progress.

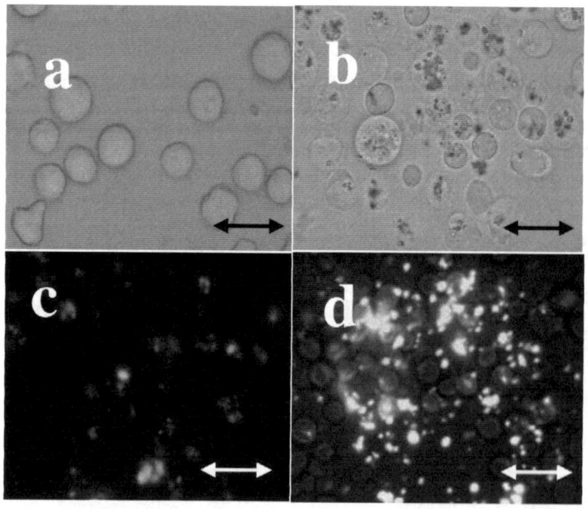

Figure 7 Live M1 cells (a) and those up-taking Y_2O_3 nanoparticles (b, c and d). (a) and (b) are bright field images; (c) and (d) are UC and NIR fluorescence images under 980-nm excitation in biological microscopic view (scale bar: 25 μm).

4. Conclusions

Er^{3+} doped Y_2O_3 ceramic nanophosphors materials were successfully prepared by enzymatic and polymer assisted homogeneous co-precipitation method. Resultant particles size is approximately 30 nm with narrow size distributions. Surface modification improves the chemical durability of nanoparticles as well as significantly reduces the agglomeration under physiological conditions. Surface modified nanoparticles generate lower toxic effect in M1 cells while compare to bare nanoparticles. After the intra cellular uptake by M1 cells nanoparticles shows green, red UC and NIR fluorescence under 980-nm excitation. Therefore, we concluded that surface modified Er^{3+} doped Y_2O_3 ceramic nanophosphors are promising materials for UC and NIR bioimaging under NIR excitation.

Acknowledgments

This work is supported by Academic Frontier Project for Private Universities: Matching Fund Subsidy from MEXT (Ministry of Education, Culture, Sports, Science and Technology, Japan), 2006-2010 and also partly from NEDO industrial research project "Development of upconversion nanoparticles for Bio-Nano-Photonics". We are grateful to Prof. T. Ishiguro for the TEM observation.

References

[1] Matsuura D 2002 *Appl. Phys. Lett.* **81** 4526

[2] Matsuura D, Ikeuchi T and Soga K 2008 *J. Lumin.* **128** 1267

[3] Hampl J, Hall M, Mufti M.A, Yao Y.M, Mac Queen D.B, Wright W.H. and Cooper D.E. 2001 *Anal. Biochem.* **288** 176

[4] Unuma H 2007 *Int. J. Appl. Ceram. Technol.* **4** 14

[5] Peng Z, Jiang Y, Song Y, Wang C and Zhang H 2008 *Chem. Mater.* **20** 3153

[6] Shevach E.M. and Rosenthal A.S. 1973 *J. Exp. Medicine.* **138** 1213

[7] Krombach F, Munzing S, Allmeling A.M, Gerlach J.T, Behr J and Dorger M. 1997 *Environ. Health. Perspect.* **105** 1261

[8] http://en.wikipedia.org/wiki/Macrophage

[9] Venkatachalam N, Saito Y and K. Soga 2009 *J. Am. Ceram. Soc.* (JACE 02986; article in press)

[10] Kamimura M, Miyamoto D, Saito Y, Soga K and Nagasaki Y 2008 *Langmuir* **24** 8864

[11] Kamimura M, Miyamoto D, Saito Y, Soga K and Nagasaki Y 2008 *J. Photopolym. Sci. Technol.* **21** 183

[12] Zako T, Nagata H, Terada N, Utsumia A, Sakono M, Yohda M, Ueda H, Soga K and Maeda M 2009 *Biochem. Biophys. Res. Comm.* **381** 54

[13] Zako T, Nagata H, Terada N, Sakono M, Soga K and Maeda M 2008 *J. Mater. Sci.* **43** 5325

Journal of Physics: Conference Series **191** (2009) 012003 doi:10.1088/1742-6596/191/1/012003

Effect of K_3PO_4 addition as sintering inhibitor during calcination of Y_2O_3 nanoparticles

K Soga[1,2], Y Okumura[1], K Tsuji[1] and N Venkatachalam[2]

[1]Department of Materials Science and Technology, Tokyo University of Science 2641 Yamazaki, Noda, Chiba 278-8510, Japan

[2]Polyscale Technology Research Center, Tokyo University of Science 2641 Yamazaki, Noda, Chiba 278-8510, Japan

E-mail: mail@ksoga.com

Abstract. Erbium-doped yttrium oxide nanoparticle is one of the most important for fluorescence bioimaging under near infrared excitation. Particle size of it below 100 nm is an important requirement for a cellular bioimaging. However, the synthesis with such small particles is difficult at the calcination temperature above 1200 °C due to the sintering and crystal growth of the particles. In this study, yttrium oxide nanoparticles with average size of 30 nm were successfully synthesized by using K_3PO_4 as a sintering inhibitor during the calcination. A single phase of cubic Y_2O_3 as the resultant material was confirmed by XRD, which was also confirmed to emit a bright upconversion emission under 980-nm excitation. Improvement of chemical durability due to the introduction of phosphate group on the surface of the Y_2O_3 particles is also reported.

1. Introduction

Fluorescence bioimaging is one of the most important tools for biological researches and medical diagnosis to visualize biological phenomena. Current problems for the fluorescence bioimaging are color fading of the imaging probes, photonic damage to the biological substances, auto fluorescence and strong scattering due to the use of short-wavelength light such as ultra violet light as an excitation light. The authors have developed fluorescence bioimaging system that can be excited by near infrared (NIR) light by using ceramic nanophosphors as fluorescence probes [1, 2]. Rare-earth-doped ceramic nanophorsphors (RED-CNP) have developed as NIR bioimaging probes. Yttrium oxide (Y_2O_3) nanoparticles doped with erbium (Er) is one of the best candidates because nanoparticles are available by various precursor precipitation routs and they can both emit visible upconversion (UC) emission in visible wavelength region and NIR emission around 1500-nm wavelength region [3-13].

An important requirement to the RED-CNP to be used for the bioimaging is sizes between 10 and 200 nm [1] to keep both of the emission intensities and the dispersion in physiological conditions. Various methods to prepare Y_2O_3 particles with 20-200 nm size have been reported to date [3-13]. For the synthesis, calcination temperatures below 900 °C were essential to keep the particles un-sintered and un-grown from the controlled size of the precursors [13]. On the other hand, residual components such as hydroxyl or carbonate

Journal of Physics: Conference Series **191** (2009) 012003 doi:10.1088/1742-6596/191/1/012003

groups are harmful to have an efficient emission from the material [14]. The removal of the residuals is still doubtful with the calcination temperature of 900 °C [15]. However, when increased with the calcination temperature above 900 °C, it should be difficult to keep the particles size to be less than 100 nm. It is required to have Y_2O_3 particles with a size less than 100 nm for the use of the particles as a imaging probe for an intra cell imaging.

The first objective of the present study is to decide a proper calcination temperature for achieving an efficient upconversion emission from Er^{3+}-doped Y_2O_3 particles. Secondly, a use of K_3PO_4 as an sintering inhibitor to prevent the sintering between Y_2O_3 particles during the high temperature calcination at 1200 °C. Some extensional effects such as chemical durability improvement are also reported.

2.Experimental

2.1. Materials and sample preparation

Erbium nitrate pentagonal hydrate ($Er(NO_3)_3$ $5H_2O$: purity 99.95%), yttrium nitrate hexagonal hydrate ($Y(NO_3)_3$ $6H_2O$: purity 99.99%) and tri-potassium phosphate n-hydrate (K_3PO_4 nH_2O : purity 80%) were supplied from Kanto Chemical Company (Japan), sodium carbonate (Na_2CO_3 : purity 99%) were supplied from High Purity Chemical Laboratory Company (Japan).

For the experiments varying calcination temperature, hydroxide precursors were prepared by using a conventional alkaline co-precipitation method. The samples were calcined in air for 5 hrs at various temperatures between 800 and 1200 °C.

Er^{3+}-doped Y_2O_3 nanoparticles were synthesized by calcining a hydroxyl carbonate precursor [5, 13]. In order to obtain the precursors, an aqueous solutions containing 200 ml of $Er(NO_3)_3$ (0.3 mmol/L) and $Y(NO_3)_3$ (4 mmol/L) were mixed into 200 ml of Na_2CO_3 (0.06 mol) aqueous solution and kept stirred for 0.5 hrs. The precipitation was washed by centrifugal washing for three times and dried at 80 °C overnight in air.

A fixed amount of $Y(OH)CO_3$ precursor was dispersed into various amount of K_3PO_4 aqueous solution with a fixed concentration so that molar ratio of Y: K_3PO_4 (denoted as 1:*x* in this paper) varies from 1:0 to 1:50. The mixture was maintained for 1 day at room temperature and dried by using a vacuum freeze dryer. The dried samples were calcined at 1200 °C for 5 hours and cooled in a furnace to room temperature. To remove K_3PO_4, particles were applied with centrifugal-wash until the pH reaches to 7 by distilled water and dried at 80 °C overnight in air.

2.2 Measurements

UC emission spectra were measured by using a spectrofluorometer (RF-5000, Shimadzu) under an excitation by IR diode laser at 980 nm. The resultant materials were supplied for XRD (XRD-6100, Shimadzu with Cu Kα radiation), FT-IR measurement (FT-IR8300, Shimadzu), SEM (S-4200, Hitachi) and TEM (LEO922, Carl Zeiss) imaging, which were performed by a . To test the chemical durability of the particles, 50 mg of the particles were at first dispersed into 50 mL NH_4OHaq with pH=7. The pH of the solution was measured after 1.0 M HCl aq. (0.2 mL) addition to make the initial solution pH to be 3.

3. Results and Discussion

Figure 1 shows the dependencies of UC emission intensity and infrared absorption due to hydroxyl groups on calcination temperatures. The values were obtained by integrating in corresponding regions in the UC emission and FT-IR spectra, respectively. By increasing the calcination temperature, the UC emission intensity increases and tends to saturate toward 1200 °C. Among the factors that control the UC emission intensity, the residual amount of the hydroxyl group affects on the upconversion intensity most seriously since the existence of the hydroxyl group heavily quenches the emission of the rare-earth ions due to the large vibration energy of the hydroxyl group (approx. 3400 cm-1). The similar behaviors of the increase of the UC intensity and the decrease of the absorption due to the

Journal of Physics: Conference Series **191** (2009) 012003 doi:10.1088/1742-6596/191/1/012003

hydroxyl group suggests that a calcination temperature above 1200 °C is required for an efficient emission from the RE-doped Y_2O_3 particles. On the other hand, it was found that the particle size of the particles calcined at 1200 °C was as large as several hundreds nm even when the calcinatinon started from the precursors with several tens nm size.

To obtain the particles with the size less than 100 nm, sintering inhibition by introducing K_3PO_4 was preformed. Figure 2 shows the XRD patterns of the calcined samples with or without K_3PO_4. It

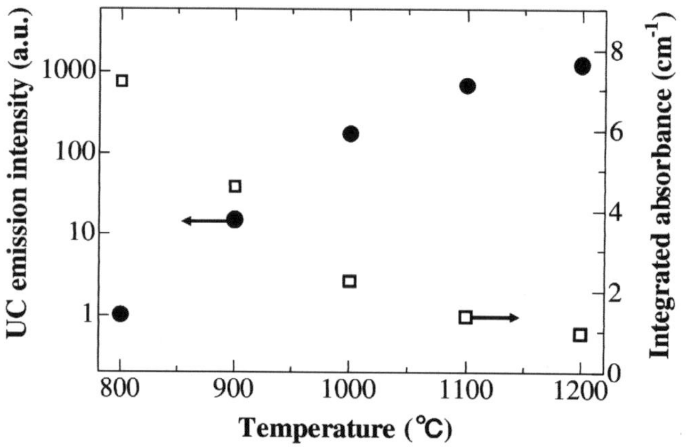

Figure 1. Dependencies of UC emission intensity (closed circle) and infrared absorption due to hydroxyl groups (open square) on calcination temperatures.

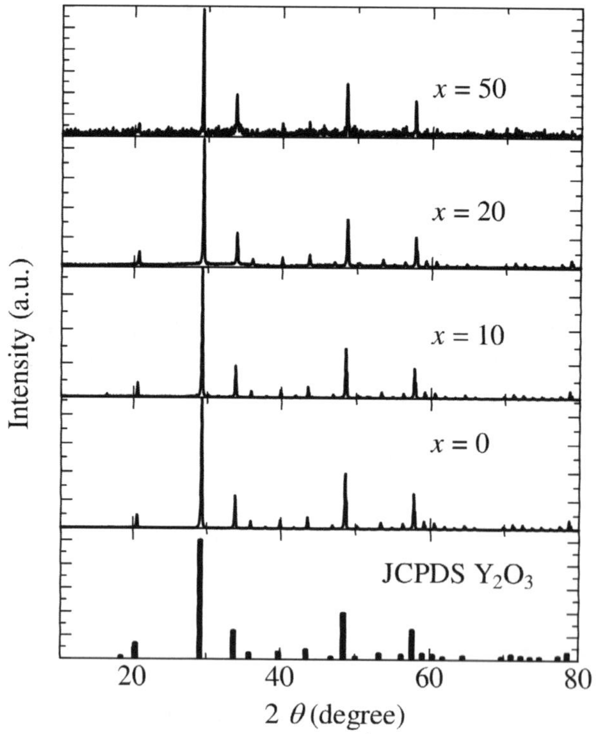

Figure 2. XRD patterns of calcined samples together with JCPDS card data. *x* value in the figure denotes the molar ration of the K_3PO_4 to Y.

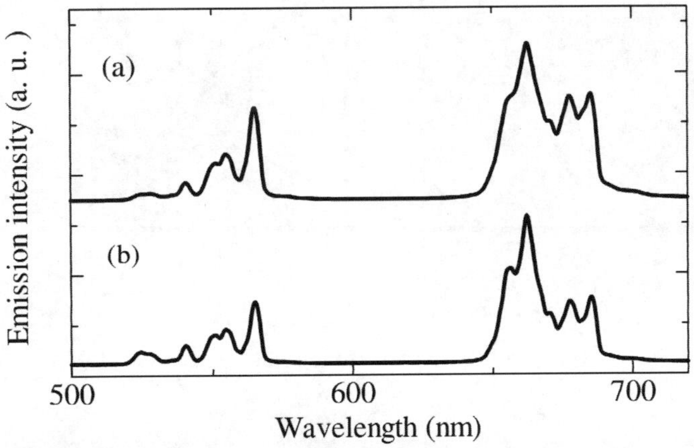

Figure 3. Upconversion emission spectra of the samples calcined without (a) or with (b) K_3PO_4 ($x = 20$).

Figure 4. SEM images of samples. The ratios in the pictures denote the molar ratios of the Y : K_3PO_4 (1 : x) during the calcination.

was found that the resultant particles are single phase cubic Y_2O_3 even in the case to calcine the precursor with 50 times larger molar amount of K_3PO_4 to Y. Figure 3 shows the UC emission spectra of Er in the samples calcined with or without K_3PO_4 under 980-nm excitation. The similarity of the spectra also supports that the resultant particles are in the same cubic Y_2O_3 as the samples without K_3PO_4 and not any of the compound contains phosphate groups.

Figure 4 shows the SEM images of the Y_2O_3 calcined with various amount of K_3PO_4. Without the addition of the K_3PO_4, the particles were sintered together to form a large agglomeration to form particles with sizes more than 1 μm. By increasing the amount of K_3PO_4, small particles with several tens nm size increased. When added with 50 times more molar ration of K_3PO_4 to Y, no large particle with a size more than 100 nm was found. The gradual increase of the small particles corresponds to the increase of K_3PO_4 is caused by the sintering inhibition effects by the K_3PO_4 among the precursor particles. According to the SEM and TEM observation of the resultant particles obtained by calcination with K_3PO_4 ($x = 50$) in Figure 5, the particle size ranged from 20 to 40 nm with an average

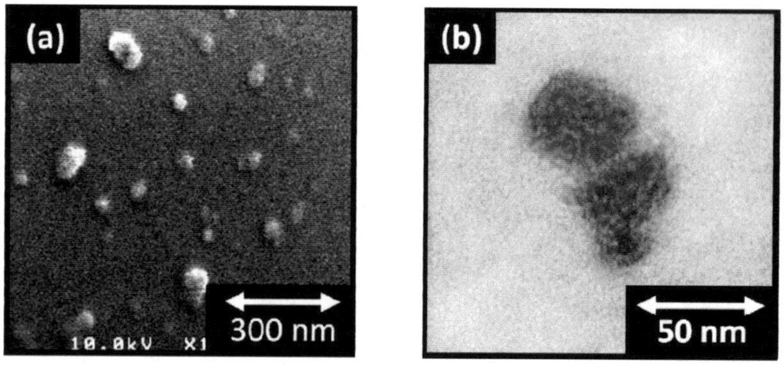

Figure 5. SEM (a) and TEM (b) images of a sample calcined with K_3PO_4 (x=50).

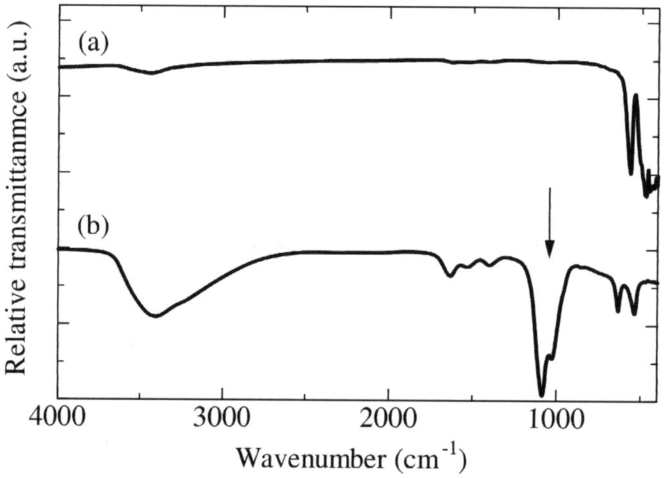

Figure 6. FT-IR spectra of the calcined samples without (a) or with (b) K_3PO_4 (x =20).

of 30 nm. Thus, we could successfully synthesized Y_2O_3 particles with size of several tens nm calcined at 1200 °C.

Figure 6 shows the FT-IR spectra of the Y_2O_3 particles calcined with or without K_3PO_4 inhibitor. Peaks round 1100 cm^{-1} denoted as an arrow in the figure corresponds to the absorption due to PO_4 group. The absorption is only observed in the spectrum of the particles calcined with K_3PO_4 inhibitor. Along with the PO_4-absorption, characteristic peaks of a stretching mode hydroxyl group at 3400 cm-1 and bending mode of water molecule around 1500 cm^{-1} were observed only in the spectrum of the particles calcined with K_3PO_4 inhibitor. As a result of the XRD and UC emission experiments, the resultant particles are mostly crystalline Y_2O_3 while the FT-IR shows the existence of the PO_4 groups and water molecules together with the particles. Summarizing the whole results, the PO_4 groups exist on the surface of the Y_2O_3 particles attracting water molecules.

Chemical durability of the particles under acidic condition was tested and the result is shown in Figure 7. It is known that Y_2O_3 particles dissolve into water under pH=6.4 [16]. The pH gradually decreases and reaches to pH=6.4 in the case of the Y_2O_3 particles without K_3PO_4 inhibitor. On the other hand, the particles with K_3PO_4 inhibitor do not change the pH of the water to be 3 even after 10 hrs. This result also support the existence of the chemically durable PO_4 groups on the surface of the Y_2O_3 particles with K_3PO_4 inhibitor. By the use of the K_3PO_4 inhibitor, chemical durability of the Y_2O_3 particles under acidic condition was improved well to be used under physiological conditions.

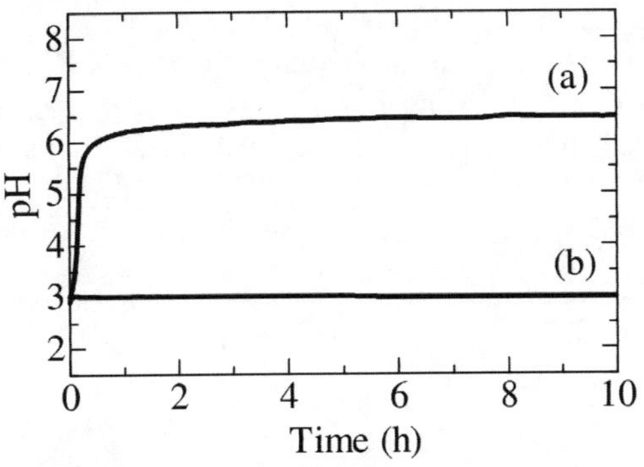

Figure 7. pH variation per hour after acid addition into water dispersed with Y_2O_3 particles calcined without (a) or with (b) K_3PO_4 ($x = 20$).

4. Conclusion

Required calcination temperature of Er-doped Y_2O_3 for efficient upconversion emission was found to be above 1200 °C, where the particle size under normal calcination grows up to be above 100 nm. By using K_3PO_4 as a sintering inhibitor, the particle size of the Y_2O_3 calcined at 1200 °C was successfully suppressed to be approximately 30 nm as average. The surface of the Y_2O_3 particles calcined with K_3PO_4 is covered with phosphate group, which improved the chemical durability of the particles under acidic conditions.

Acknowledgement

This work was supported by "Academic Frontier" Project for Private Universities: Matching Fund Subsidy from MEXT (Ministry of Education, Culture, Sports, Science and Technology), 2006-2010.

References

[1] Soga K, Tsuji T, Tashiro F, Chiba J, Oishi M, Yoshimoto K, Nagasaki Y, Kitano K and Mamaguchi S 2008 *J. Phys. Conf. Ser* **106** 012023
[2] Konishi T, Shimizu M, Kameyama Y and Soga K 2007 *J. Mater. Sci. Mater. in Electron.* **18** S183
[3] Zako T, Nagata N, Terada N, Utsumi A, Sakono M, Yohda M, Ueda H, Soga K and Maeda N, *Biochem. Biophys. Res. Commun.* 2009 **381** 54
[4] Kammimura M, Miyamoto D, Saito Y, Soga K and Nagasaki Y 2008 *Trans. Mater. Res. Soc. Jpn.* **33** 795
[5] Saito Y, Shimizu K, Kamimura M, Furusho H, Soga K and Nagasaki Y 2008 *Trans. Mater. Res. Soc. Jpn.* **33** 803
[6] Matsuura D, Ikeuchi T and Soga K 2008 *J. Lumin.* **128** 1267
[7] Konishi T, Shimizu K, Saito Y and Soga K 2007 *J. Photopolym. Sci. and Tech.* **20** 11
[8] Konishi T, Yamada M, Soga K, Matsuura D and Nagasaki Y 2006 *J. Photopolym. Sci. and Tech.* **19** 145
[9] Soga K, Koizumi R, Yamada M, Matsuura D and Nagasaki Y 2005 *J. Photopolym. Sci. and Tech.* **18** 73
[10] Zako T, Nagata H, Terada N, Sakono M, Soga K and Maeda M 2008 *J. Mater. Sci.* **45** 5325
[11] Kamimura M, Miyamoto D, Saito Y, Soga K and Nagasaki Y 2008 *J. Photopolym. Sci. and Tech.* **21** 183

[12] Kamimura M, Miyamoto D, Saito Y, Soga K and Nagasaki Y 2008 *Langmuir* **24** 8864
[13] Venkatachalam N, Saito Y and Soga K, 2009 *J. Am. Ceram. Soc.* **92** 1006
[14] Soga K, Wenzhong W, Riman R E, Bryan J B and Mikeska K R 2003 *J. Appl. Phys.* **93** 2946
[15] Zhang J, Tang Z, Zhang Z, Fu W, Wang J and Lin Y 2002 *Mater. Sci. and Eng.* **A334** 246
[16] Pourbaix M, 1974 Atlas of Electrochemical Equilibria in Aqueous Solutions, NACE, Houston

Journal of Physics: Conference Series **191** (2009) 012004 doi:10.1088/1742-6596/191/1/012004

Random Walk Analysis of Grain Motion during Superplastic Deformation of TZP

T Okamoto[1], K Yasuda[1] and T Shiota[1]

[1]Department of Metallurgy and Ceramics Science, Graduate School of Science and Engineering, Tokyo Institute of Technology, 2-12-1-S7-14, Ookayama, Meguro-ku, Tokyo, Japan

E-mail: okamoto.t.ag@m.titech.ac.jp

Abstract. This study focuses on grain motion in TZP (Tetragonal Zirconia Polycrystal) ceramics during superplastic deformation. The specimen was 16 times elongated repeatedly at 1400℃ in air. The increment of true plastic strain was set to be 2%, and the specimen was deformed up to 30.3% true plastic strain finally. After each deformation, displacement vectors of specified 748 grains were measured from their position vectors determined by FE-SEM micrographs. As a result, the grains move to the tensile loading direction in zigzag way. And also, the zigzag motion changes with plastic strain: The grains move randomly (random walk motion) by the first 15% true plastic strain, and then grain motion becomes spatially uniform gradually. It is related to changes of constraint of surrounding matrix.

1. Introduction

Since superplasticity of ceramics was discovered by Wakai et al. in 1986 [1], many ceramists have studied to understand its deformation mechanism. Most studies have been discussed by means of a traditional method based on a high-temperature creep theory. However, the present authors focus on grain motion during superplastic deformation [2-4]. In this paper, superplastic deformation of a TZP (Tetragonal Zirconia Polycrystal) specimen was repeatedly conducted at 1400℃ in tension. After each deformation, displacement vectors of specified 748 grains were measured from their position vectors determined by FE-SEM micrographs. Then, grain motion was discussed by using random walk theory.

2. Experimental procedure

2.1. Preparation of TZP specimen

Raw material was ZrO_2 powder TZ-3Y (Tosoh, Japan) containing 3 mol% Y_2O_3 as a stabilization agent. This powder was uniaxially pressed at 30 MPa for 2 min, and isostatically cold-pressed at 100 MPa for 2 min. The compact was sintered at 1500℃ for 2 hr in air. The sintered body was ground and cut to obtain a tensile specimen. A gage section of the specimen was 10 mm in length and 2 mm square in cross section. Finally, the specimen surface was polished with diamond slurry(2 μm- 4 μm), and thermally etched at 1300℃ for 10 min in air.

2.2. Repeated superplastic deformation

After the specified 748 grains were observed by FE-SEM, tensile deformation was conducted at 1400℃. Crosshead speed was 0.2 mm/min (its strain rate was on the order of 1×10^{-4} [s^{-1}]). The increment of true plastic strain was set to be about 2%. After the deformation, the same grains were observed by FE-SEM again. These observations and deformations had been repeated 16 times, and the specimen was elongated to 30.3% true plastic strain finally.

3. Result and discussion

3.1. True stress/ true strain curve

Figure 1 shows a true stress/true strain curve of these deformations. All the plastic deformations seem to start at about 30 MPa. Maximum stresses are about 40 MPa at the first 4 deformations. After the 5th deformation, however, maximum stresses drop to about 35 MPa (except for the 9th deformation). Namely, the specimen was deformed easier after the 5th deformation.

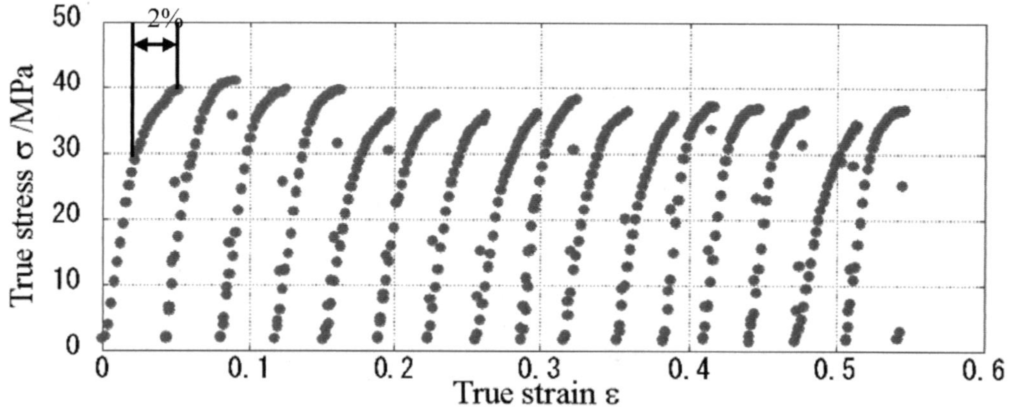

Figure 1. True stress/ true strain curve.

3.2. Displacement vectors

3.2.1. Calculation of displacement vectors of the grains

As shown in figure 2, x-axis is set to be a tensile loading direction, and y-axis is perpendicular to x-axis. Grain position vectors were determined as follows: (i) A longest line parallel to y-axis was drawn on a grain, then a midpoint of the line was set to be a representative position of the grain. (ii) An arbitrary grain was set to be a temporary origin of the coordinate. (iii) From the temporary origin, temporary position vectors of the 748 grains were calculated. (iv) A center of gravity of the grains was determined from the temporary position vectors of the grains. (v) The origin was changed to this center of gravity, and position vectors of the grains were recalculated. From equation (1), displacement vectors $\boldsymbol{u}$ were calculated,

$$\boldsymbol{u} = (x_2 - x_1, y_2 - y_1) \qquad (1)$$

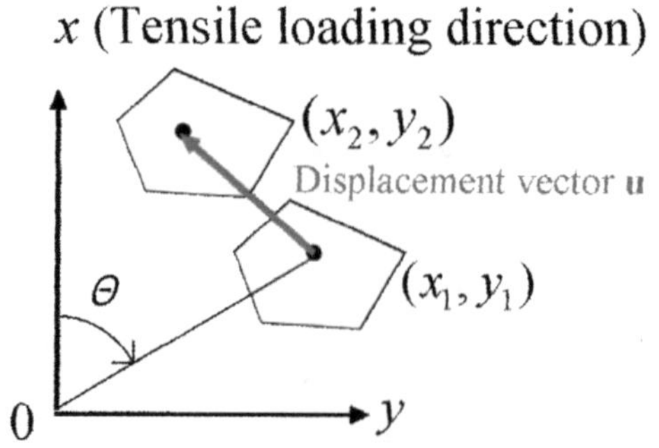

Figure 2. Measurement of a displacement vector.

Journal of Physics: Conference Series **191** (2009) 012004 doi:10.1088/1742-6596/191/1/012004

where (x_1, y_1) is a position vector of a grain before deformation, and (x_2, y_2) is that after deformation.

3.2.2. Displacement vectors map

Figure 3 shows displacement vector maps at each deformation. At the initial deformation shown in figure 3(a), the grains move to the tensile directions. However, at the second deformation, the grains move convergent to the origin though the specimen was actually elongated with 2.2% of true plastic strain increment during loading (figure 3(b)). Then, the grains move divergent against the origin again (figure 3(c)). Namely, grain movement repeats divergent out and convergent back.

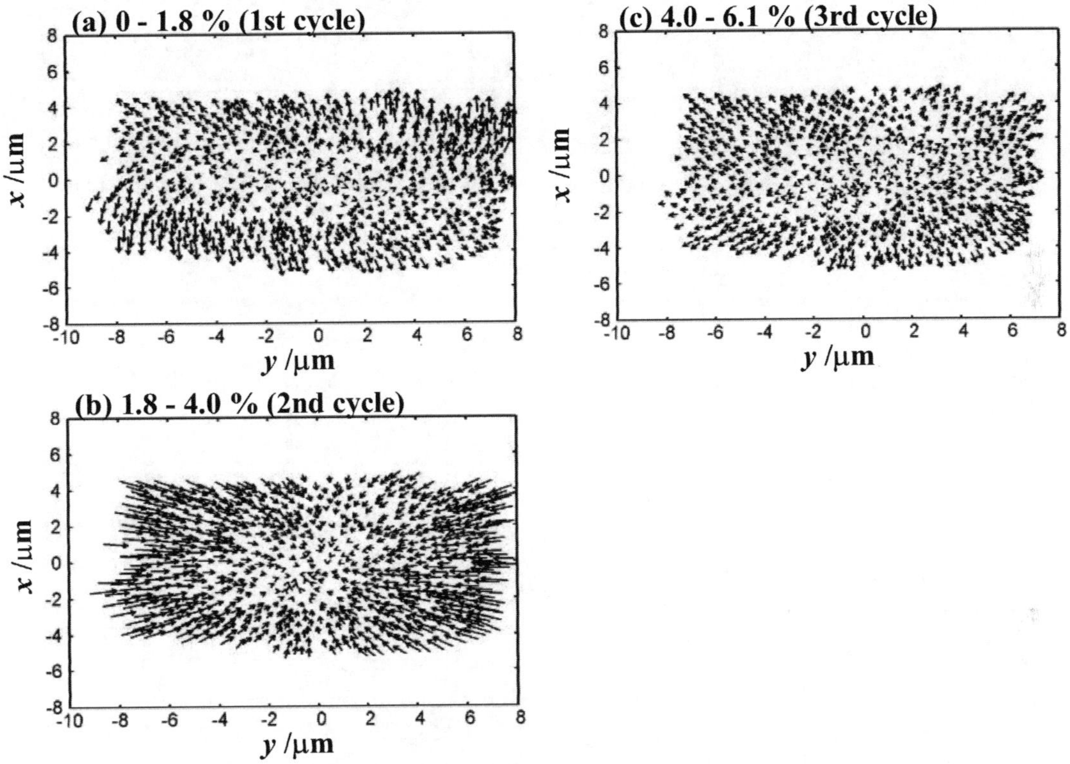

Figure 3. Displacement vector maps.

Figure 4 shows a trajectory of a grain whose initial position vector is (x_0, y_0) = (2.7 μm, 4.1 μm). We see the grain moves with zigzag way. We also note that the grain moves with smaller displacement magnitude at latter half of the deformations (i.e. after about 15% true plastic strain). If the zigzag motion is concerned with constraint of surrounding matrix, these results indicate that spatial distribution of constraint becomes uniform as superplastic deformation advances. This motion looks like random walk of Brownian motion or diffusion phenomenon.

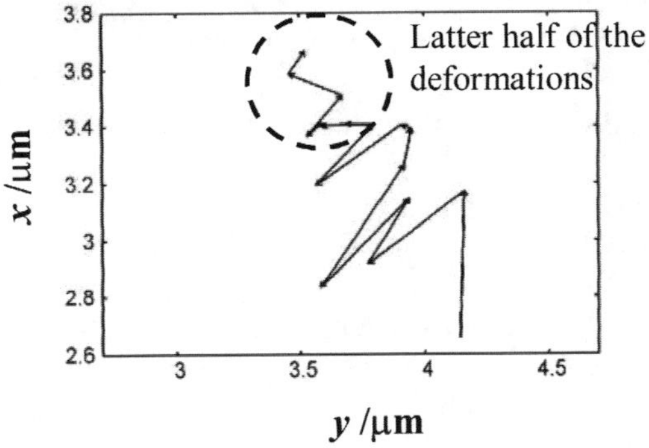

Figure 4. Trajectory of a grain.

3.2.3. Trajectory analysis of grains by random walk theory

In this section, grain motion is analyzed approximately by means of random walk theory containing drift motion. In this analysis, random walk motion is corresponding to constraint of surrounding matrix and drift motion is driven by an external stress. Namely, grain motion can be approximated by,

$$\boldsymbol{u}(\varepsilon_p) = \boldsymbol{A}\varepsilon_p + \boldsymbol{Z}(\varepsilon_p) \qquad (2)$$

where ε_p is true plastic strain, $\boldsymbol{u}(\varepsilon_p)$ is a displacement vector at ε_p, $\boldsymbol{A}$ is a constant vector along linear drift motion and $\boldsymbol{Z}(\varepsilon_p)$ is a random walk vector. Drift motion of the grains is set to be a linear vector from 0% to 30.3% of ε_p (figure 5). Figure 6 is histograms of random walk motion at each deformation. The distribution of them becomes narrow gradually. That is, constraint of surrounding matrix becomes small as superplastic deformation advances.

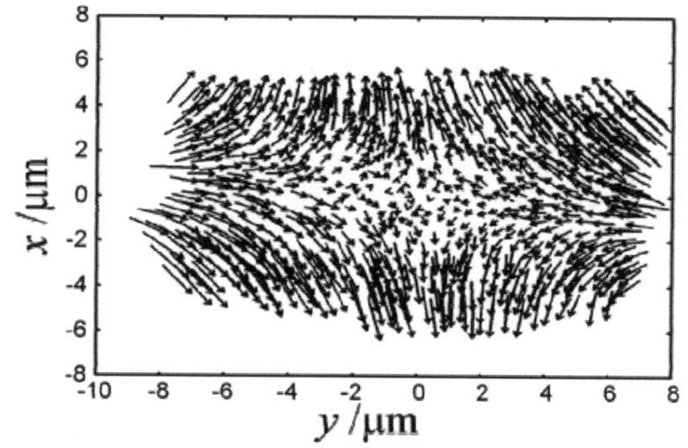

Figure 5. Drift motion of the grains.

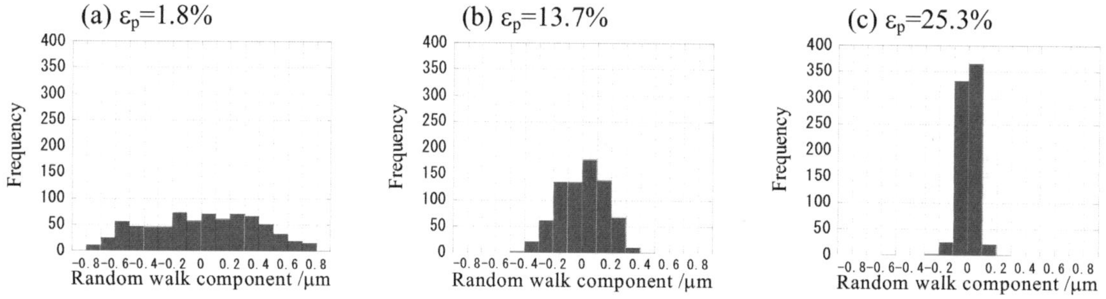

Figure 6. Histograms of random walk motion of the grains.

4. Conclusion

Superplastic deformation of the TZP specimen was repeatedly conducted. After each deformation, the displacement vectors of the specified 748 grains were measured from their position vectors determined by FE-SEM micrographs. As a result, the grains move to the tensile loading direction in zigzag way. And also the random walk analysis shows that constraint of surrounding matrix becomes small gradually as superplastic deformation advances.

References

[1] Wakai F, Sakaguchi S and Matsuno Y 1986 *Adv. Ceram. Mater.* **1** 259
[2] Yasuda K, Okamoto T, Shiota T and Matsuo Y 2006 *Mater. Sci. Eng. A* **418** 115
[3] Okamoto T, Yasuda K, Shiota T and Matsuo Y 2006 *Trans. Japan Soc. Mech. Eng. A* **72** 1080
[4] Yasuda K, Okamoto T, Shiota T and Matsuo Y 2008 *J. Solid Mechanics and Materials Engineering* **2** 1497

Journal of Physics: Conference Series **191** (2009) 012005 doi:10.1088/1742-6596/191/1/012005

Detection of Damage Initiation and its Evolution in Insulating Brick from Non-linearity of Stress/Strain Curve

R Zhang[1], K Yasuda[1], R Furushima[1] and T Shiota[1]

[1]Department of Metallurgy and Ceramics Science, Graduate School of Science and Engineering, Tokyo Institute of Technology, 2-12-1-S7-14, Ookayama, Meguro-ku, Tokyo, Japan

E-mail: kyasuda@ceram.titech.ac.jp

Abstract. This paper deals with experimental verification of the present author's theory relating to estimation of damage accumulation in materials showing non-linear stress/strain curves. The theory is based on parallel elements model and gives that the second order derivative of a stress/strain curve is proportional to a probability density function of strain at which each element is damaged. Uni-axial compression test was conducted for an insulating brick and its probability density function is obtained from the non-linear stress/strain curve. The function coincides well with acoustic emission data simultaneously measured in the compression test. This fact shows that the theory is useful for quick and brief estimation of damage initiation and its evolution in insulating bricks.

1. Introduction

Ceramic materials are usually fractured in a catastrophic manner after linear elastic deformation, however, non-linear stress/strain relationship also can be seen in some materials such as polycrystalline graphite, refractories and porous materials [1-4]. Up to now, we have attributed the non-linearity to a local damage process in the materials, and therefore measured acoustic emission during a loading test in order to reveal its damage process [3,4]. Acoustic emission is certainly an excellent technique to detect damage events in materials, however we sometimes encounter the difficulties in fixing expensive measuring devices and adjusting optimal experimental conditions. Thus, for a long time, we have wanted a quick and brief technique to estimate damage initiation and its evolution in the materials.

In this paper, we develop a theory relating to estimation of damage accumulation in materials from non-linearity in stress/strain curves, and then verify the theory with an insulating brick under compression tests.

2. Theory

Consider that uni-axial loading is given to the parallel elements model, as shown in Figure 1. The loading is carried out under a constant displacement rate. Each element has its own damage strain depending on the probability density function $f(\varepsilon)$ (viz. $f(\varepsilon)d\varepsilon$ means the probability that an element is damaged at the strain from ε to $\varepsilon+d\varepsilon$). The initial Young's modulus of all elements is denoted by E, and that after damaged is E'. It follows that the constitutive equation of an element j can be expressed as below [5,6],

$$\begin{cases} \sigma = E\varepsilon & (\varepsilon < \varepsilon_{dj}) \\ \sigma = \sigma_{dj} = E\varepsilon_{dj} & (\varepsilon = \varepsilon_{dj}) \\ \sigma = E'(\varepsilon - \varepsilon_{dj}) + \sigma_{dj} & \\ \quad = E'\varepsilon + (E - E')\varepsilon_{dj} & (\varepsilon > \varepsilon_{dj}) \end{cases} \quad (1)$$

,where σ_{dj} and ε_{dj} are the damage stress and strain of the element j, respectively. For simplicity, we begin with finite number of element (viz. j = 1~n). In advance, ε_{dj} may be renumbered in an ascendant order. When the applied strain ε is in the region of $0 \le \varepsilon \le \varepsilon_{d1}$, any element is not damaged, and the applied stress σ of the model can be expressed as below,

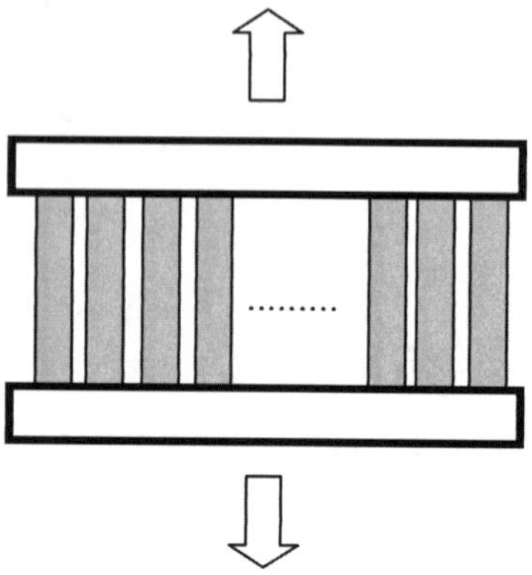

Figure1. Parallel elements model.

$$\sigma = \sum_{j=1}^{n} V_j E\varepsilon = E\varepsilon \quad (2)$$

,where V_j is the volume fraction of the element j. When the applied strain ε reaches ε_{d1}, the first element has just got damaged and then the applied stress is,

$$\sigma = V_1\sigma_{d1} + \sum_{j=2}^{n} V_j E\varepsilon = V_1\sigma_{d1} + (1 - V_1)E\varepsilon \quad (3)$$

Furthermore, the applied strain ε is in the region of $\varepsilon_{d1} \le \varepsilon \le \varepsilon_{d2}$, the applied stress is,

$$\sigma = V_1\{E'\varepsilon + (E - E')\varepsilon_{d1}\} + (1 - V_1)E\varepsilon \quad (4)$$

With increasing the applied strain, each element is damaged one by one, and therefore the overall Young's modulus of the material gradually decreases according to the equations below,

$$\sigma = \sum_{j=1}^{k-1} V_j E'\varepsilon + \sum_{j=1}^{k-1} V_j (E - E')\varepsilon_{dj} + V_j\sigma_{dj} + \left(1 - \sum_{j=1}^{k} V_j\right)E\varepsilon \quad (\varepsilon = \varepsilon_{dk}) \quad (5)$$

$$\sigma = \sum_{j=1}^{k} V_j E'\varepsilon + \sum_{j=1}^{k} V_j (E - E')\varepsilon_{dj} + \left(1 - \sum_{j=1}^{k} V_j\right)E\varepsilon \quad (\varepsilon_{dk} < \varepsilon < \varepsilon_{d(k+1)}) \quad (6)$$

If the number n of element becomes infinite, V_j is approaching to infinitely small, and consequently the equation (5) is identical to the equation (6). By using the probability density function $f(\varepsilon)$ of the damage strain (viz. $\sum_{j=1}^{k} V_j$ is replaced by $\int_0^{\varepsilon} f(t)dt$), we obtain the non-linear stress/strain relationship by the following equation[5,6].

$$\sigma = E'\varepsilon \int_0^{\varepsilon} f(t)dt + (E - E')\int_0^{\varepsilon} f(t)tdt + E\varepsilon\left[1 - \int_0^{\varepsilon} f(t)dt\right] \quad (7)$$

After some differential operations of equation (7), we eventually obtain the following relation.

$$f(\varepsilon) = \frac{1}{E - E'}\left[-\frac{d^2\sigma}{d\varepsilon^2}\right] \qquad (8)$$

Equation (8) means that the second order derivative of a stress/strain curve is proportional to a probability density function of damage strain $f(\varepsilon)$.

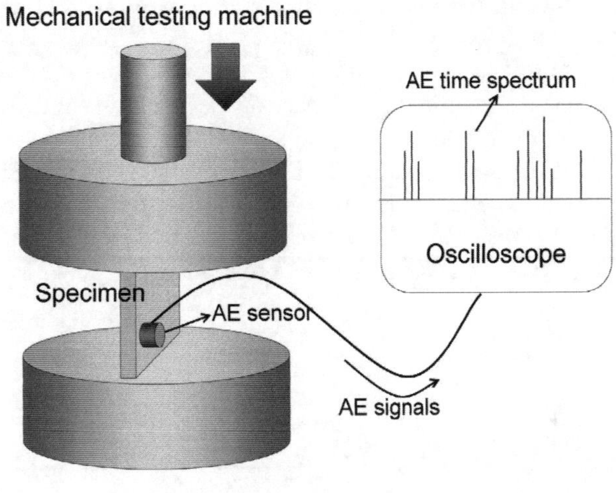

Figure 2. Uni-axial compression test.

3. Experimental procedure

3.1. Specimen

To verify the equation (8), an insulating brick was used as a typical material showing non-linear stress/strain curve. Plate-like specimens were cut out from the insulating brick (Isolite Insulating Products Co., Ltd., type C1). Its dimensions were 50mm in width by 14mm in thickness by 65mm height. The porosity was estimated to be 57.8% from the bulk density.

3.2. Uni-axial compression test

As shown in Figure 2, uni-axial compression tests were carried out with a constant crosshead speed of 0.1mm/min. During the loading, load and displacement data were collected by a mechanical testing machine (JT-Tohshi, type SC-100H). At the same time, acoustic emission was detected by AE sensor (NF Electronic Instruments, AE-90IS) attached to the surface of the specimen, and recorded by oscilloscope (Yokogawa Electric, DL7440) to obtain AE time series data.

In addition to single compression test, cyclic compression tests (viz. loading/unloading cycles) were also conducted to see how the damage accumulation was in the specimen.

4. Result and discussion

4.1. Single compression test

Figure 3(a) shows a stress/strain curve of the specimen. We see that non-linearity (convex upward) begins at the strain around 0.0025. Although we find another non-linearity (concave upward) at the beginning of loading, this is caused by unavoidable

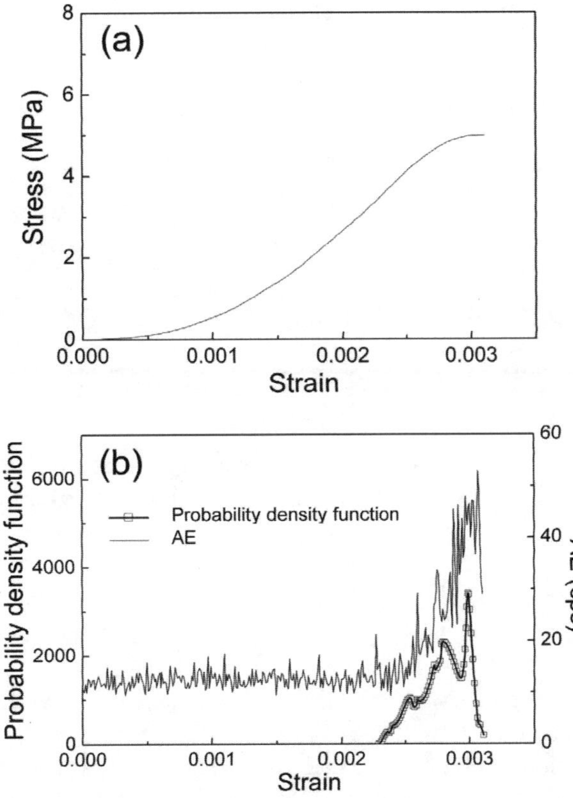

Figure 3. Single compression test, (a) stress/strain curve, (b) $f(\varepsilon)$ and AE.

clearance in the loading train. To eliminate this additional displacement, the stress/strain curve of the specimen is calibrated by that of large steel block. By using the equation (2), the probability density function of damage strain $f(\varepsilon)$ is obtained from the calibrated stress/strain curve, as shown in Figure 3(b). The function $f(\varepsilon)$ appears at the strain around 0.0023, and increases linearly, and finally has a maximum at 0.0030.

The acoustic emission (AE) data is also shown in Figure 3(b). The count rate of AE keeps constant at first; however, it abruptly increases at the strain around 0.0023, and then reaches the maximum at 0.0030. This change is almost the same as that in $f(\varepsilon)$, indicating that damage estimation from non-linearity in stress/strain curves is effective .

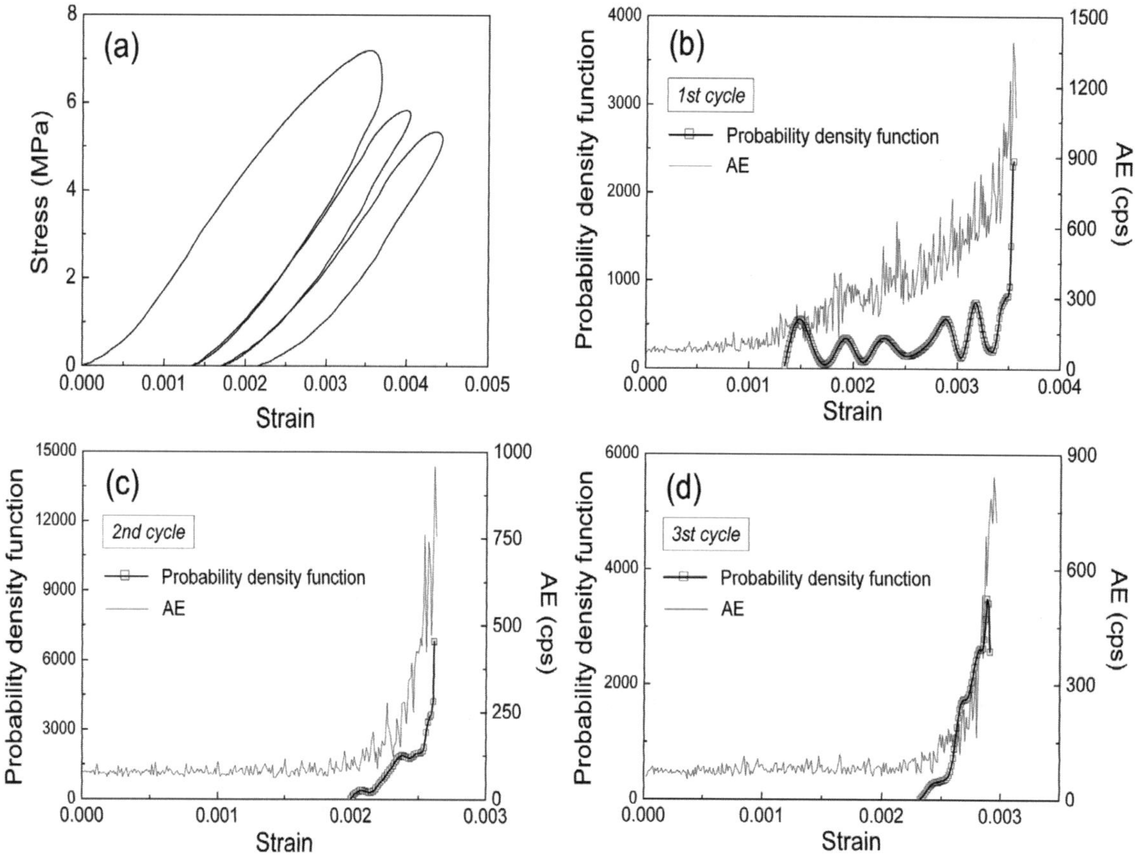

Figure 4. Cyclic compression test, (a) stress/strain curves, (b) $f(\varepsilon)$ and AE in the 1st cycle, (c) $f(\varepsilon)$ and AE in the 2nd cycle, (d) $f(\varepsilon)$ and AE in the 3rdcycle.

4.2. Cyclic compression tests

Figure 4(a) shows stress/strain curves of the specimen in cyclic compression tests. The loading/unloading cycles are three in this case.

Figure 4(b), (c), and (d) show comparison between $f(\varepsilon)$ and AE in the 1st, 2nd, and 3rd cycle, respectively. From these figures, we can see that the probability density function of damage strain $f(\varepsilon)$ coincides well with AE data in each cycle. This fact indicates that non-linear analysis of stress/strain curves provides quantitative information of damage initiation and subsequent change of the materials, equivalent to AE measurement.

It should be noted that damage initiation in Figure 3(b) is slightly different from Figure 4(b). This must be caused by the sampling problem because the thickness of specimens is not fully large enough to the size of microstructure. However, in this study, the authors dare to use thin specimen because clear AE data should be collected. If we use thicker specimens, we must obtain the same strain at initiation. So, we need more works to do in future, however, the fundamentals to detect damage initiation and its evolution from stress/strain curves of materials has just been shown in this study.

5. Conclusion

Uni-axial compression tests were carried out to an insulating brick to verify the present author's theory relating to estimation of damage accumulation in materials showing non-linear stress/strain curves. Parallel elements model brings that the second order derivative of a stress/strain curve is proportional to a probability density function of strain at which each element is damaged. From the actual stress/strain curve of the insulating brick, its probability density function is gained. The function coincides well with acoustic emission data simultaneously measured in the compression test. This result shows that the theory is useful for quick and brief estimation of damage initiation and its evolution in insulating bricks.

Acknowledgement

The author greatly express their gratitude to Dr. T. Otsuka and Dr. K. Siono of Isolite Insulating Products Co., Ltd. for their materials supply.

References

[1] Gogotsi G A and Drozdov A V 1986 *Refractories and Industrial Ceramics*, **27** 200
[2] Gogotsi G A 1988 *Powder Metallurgy and Metal Ceramics*, **27** 908
[3] Ioka I, Yoda S and Konishi T 1990 *Carbon* **28** 489
[4] Schmitt N, Berthaud Y and Poirier J 2000 *J. Euro. Ceram.Soc.* **20** 2239
[5] Yasuda K, Furushima R, Matsuo Y and Shiota T 2006 *Abs.Fall Meet.Ceram.Soc.Jpn.* (Kofu, Sep.19-21, 2006) **19** 127
[6] Yasuda K, Furushima R, Matsuo Y and Shiota T, 2007 *Abs.Fall Meet.Ceram.Soc.Jpn.* (Nagoya, Sep.12-14, 2007) **20** 109

Journal of Physics: Conference Series **191** (2009) 012006 doi:10.1088/1742-6596/191/1/012006

Estimation of Weibull parameters from parameters of initial distribution of flaw size

C Wakabayashi[1], K Yasuda[1] and T Shiota[1]

[1]Department of Metallurgy and Ceramics Science, Graduate School of Science and Engineering, Tokyo Institute of Technology, 2-12-1-S7-14, Ookayama, Meguro-ku, Tokyo, Japan

E-mail: wakabayashi.c.aa@m.titech.ac.jp

Abstract. The distribution of the largest flaw size is derived from the initial distribution of flaw size based on extreme value statistics, and also the distribution of fracture origin size is given by transforming Weibull distribution by fracture mechanical relation. These two distributions are equivalent under uniaxial loading. By using this relation, their parameters are related each other and Weibull parameters are estimated from the parameters of the initial distribution of flaw size and the number of links.

1. Introduction

In ceramics processing, flaws such as pores, large grains and cracks are generated. And they lead to material fracture. The flaws' shape and size are not uniform and the fracture strength shows wide scattering. Therefore, in order to use ceramics as structural components, reliability analysis is essential. For this purpose, fracture strength test is conducted with many specimens. And it is assumed that the fracture strength follows Weibull distribution and Weibull parameters are estimated. However, this method takes long time and low cost-efficiency, so that more economical way including non-destructive inspection is in need of reliability analysis.

The authors derived the exact distribution of fracture strength from the initial distribution of flaw size via the distribution of the largest flaw size. [1-3] Parameters of the distribution of the largest flaw size are parameters of the initial distribution of flaw size and the number of links. On the other hand, Matsuo et al. [4] obtained the distribution of fracture origin size by transforming Weibull distribution by fracture mechanical relation. These two distributions are equivalent. Therefore, the Weibull parameters are related to the parameters of the initial distribution. By using only the parameters of the initial distribution of flaw size, the distribution of fracture strength can be estimated theoretically. Due to this method, the low cost efficiency of the current reliability analysis method is solved.

2. Theory

2.1. Derivation of the exact distribution of fracture strength based on the initial distribution of flaw-size

In extreme value statistics, there are two kinds of the initial distribution. The difference is convergence of the probability. The probability of exponential type rapidly decreases when the random valuable increases. On the other hand, Cauchy type's convergence is comparatively slower. In general, the number of small flaw is far larger than the number of large flaw in brittle materials. Therefore, the initial distribution of flaw size should be exponential type distribution. Gamma distribution, one of exponential type distribution, is widely used for its flexible shape change by their parameters. In this paper, gamma distribution is used as the initial distribution of flaw size Φ_1. The probability density function of gamma distribution is given as follows,

$$\phi_1(x) = \frac{1}{\Gamma(\upsilon)}\alpha^{\upsilon}x^{\upsilon-1}\exp(-\alpha x) \quad (1)$$

, where x is flaw size, α and ν are scale and shape parameters, respectively. The validation of gamma distribution for the initial distribution of flaw size is demonstrated by applying flaw size data to Eq. (1). These data were reported by Zhang et al. [5] and Abe et al. [6]. Hereafter, Zhang's specimens are denoted by AL1 and Abe's are AL2, AL3 and AL4, respectively. In Table 1, characteristics of specimens are shown. The parameters of gamma distribution are estimated by the most likelihood method and shown in Table 2. In Fig.1, the flaw size data and gamma distribution of AL1 are shown. The correlation coefficient R is 0.999 in the data. From its quite high correlation in Fig.1, gamma distribution can be used as the initial distribution of flaw size. For other specimen, AL2, AL3 and AL4, almost the same tendencies are obtained.

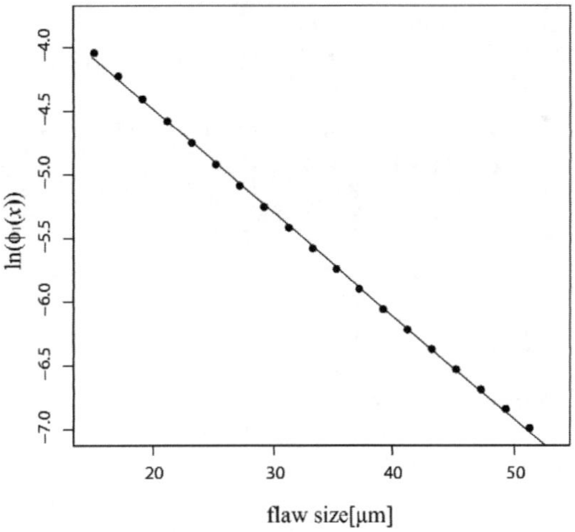

Fig.1 Flaw size data and the initial distribution of flaw size.

Table 1 Mechanical properties of alumina ceramics[4,5].

Sample	Average Strength $\overline{\sigma}$ / MPa	Shape Parameter of Weibull Distribution m / 1	Fracture Toughness K_{IC} / MPa√m
AL1	360	16.2	4.00
AL2	260	28	4.31
AL3	315	25	3.90
AL4	359	33	4.41

Table 2 Best estimators of $\hat{\alpha}$ and $\hat{\nu}$ of gamma distribution of flaw size in alumina ceramics.

Sample	$\hat{\alpha}$ / (μm)$^{-1}$	$\hat{\nu}$
AL1	0.667	14.2
AL2	0.206	5.34
AL3	0.400	9.17
AL4	0.489	11.3

The distribution of the largest flaw size is obtained from the initial distribution of flaw size based on extreme value statistics.

$$\Phi_n(x) = \left(\int_0^x \frac{1}{\Gamma(\upsilon)}\alpha^{\upsilon}x'^{\upsilon-1}\exp(-\alpha x')dx'\right)^n \quad (2)$$

Journal of Physics: Conference Series **191** (2009) 012006 doi:10.1088/1742-6596/191/1/012006

, where x is the largest flaw size, α and v are scale and shape parameters of the initial distribution of flaw size and n is the number of links in the weakest link theory. In Fig. 2, the change in the distribution of the largest flaw size of AL1 is shown. The distribution of the largest flaw size is equal to the distribution of fracture origin size under uniaxial loading. Therefore, the distribution of fracture strength is obtained by transformation of the distribution of the largest flaw size. For this transformation, the following relation in fracture mechanics is used.

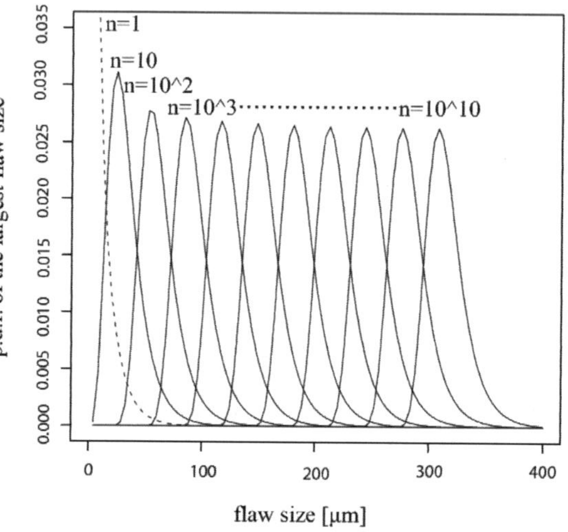

Fig.2 The largest distribution of flaw size

$$\sigma = K_{IC}\sqrt{\frac{\pi}{2c}} \qquad (3)$$

, where c is fracture origin size and K_{IC} is fracture toughness and σ is fracture strength. By using Eq. (3), the distribution of the largest flaw size is transformed to the distribution of fracture strength F_n.

$$F_n(\sigma) = 1 - \Phi_n\left(\frac{\pi K_{IC}^2}{2\sigma^2}\right) \qquad (4)$$

In order to determine the number of links, the distribution of fracture strength F_n is drawn in Fig.3, that is, in Weibull plot (x-axis is logarithm of fracture strength, and y-axis is double logarithm of $1/1\text{-}F_n$). By comparing with the Weibull parameters estimated from fracture strength data, the number of links is determined as 10^6.

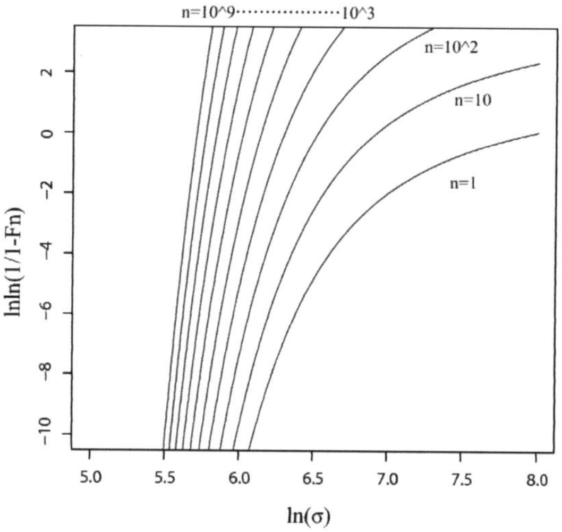

Fig.3 Weibull plot of the exact distribution of fracture strength $F_n(\sigma)$.

2.2 The distribution of fracture origin size

In former section, it is shown that the distribution of the largest flaw size is the distribution of fracture origin size. Matsuo et al. [4] obtained the distribution of fracture origin size $H(x)$ from transforming Weibull distribution by fracture mechanics. Matsuo et al. assumed Weibull distribution $F(\sigma)$ as follows,

$$F(\sigma) = 1 - \exp\left\{-\left(\frac{\sigma}{\sigma_0}\right)^m\right\} \qquad (5)$$

, where m and σ_0 are shape and scale parameters of Weibull distribution and K_{IC} is fracture toughness. By using Eq. (3) in fracture mechanics, $H(x)$ is given as below,

$$H(x) = \exp\left\{-\left(\sqrt{\frac{\pi}{2}}\frac{K_{IC}}{\sigma_0}\right)x^{-\frac{m}{2}}\right\} \quad (6)$$

This is the distribution of fracture origin size under uniaxial loading. It is originally equal to the distribution of the largest flaw size.

2.3 Method of estimation of Weibull parameters from the initial distribution of flaw size

By using the relation between the distribution of the largest flaw size $\Phi_n(x)$ and the distribution of fracture origin size $H(x)$, Weibull parameters are estimated from parameters of the initial distribution of flaw size. The parameters (α and v) in Eq. (2) are known, and the probability for any flaw size x can be calculated. On the other hand, the distribution of fracture origin size $H(x)$ has two parameters (m and σ_0). Therefore, if there are two relations including these four parameters, the parameters (m and σ_0) are determined uniquely. For this purpose, 50 percentile and 25 percentile of Eq. (2) are calculated and it is supposed that Eq.(6) gives the same probabilities. Practically, next equations are solved.

$$\Phi_n(x) = \left(\int_0^x \frac{1}{\Gamma(v)}\alpha^v x'^{v-1}\exp(-\alpha x')dx'\right)^n = 0.5 \quad (7)$$

$$\Phi_n(x) = \left(\int_0^x \frac{1}{\Gamma(v)}\alpha^v x'^{v-1}\exp(-\alpha x')dx'\right)^n = 0.25 \quad (8)$$

As the number of links, 10^6 is used to calculate 50 percentile and 25 percentile of Eq. (7) and (8) [3]. Therefore, x satisfies

$$H(x) = \exp\left\{-\left(\sqrt{\frac{\pi}{2}}\frac{K_{IC}}{\sigma_0}\right)x^{-\frac{m}{2}}\right\} = 0.5$$

$$\Leftrightarrow x = \left\{\frac{\ln 2}{V\left(\sqrt{\frac{\pi}{2}}\frac{K_{IC}}{\sigma_0}\right)^m}\right\}^{-\frac{2}{m}} \quad (9)$$

and

$$H(x) = \exp\left\{-\left(\sqrt{\frac{\pi}{2}}\frac{K_{IC}}{\sigma_0}\right)x^{-\frac{m}{2}}\right\} = 0.25$$

$$\Leftrightarrow x = \left\{\frac{2\ln 2}{V\left(\sqrt{\frac{\pi}{2}}\frac{K_{IC}}{\sigma_0}\right)^m}\right\}^{-\frac{2}{m}} \qquad (10).$$

Finally, two relations about m and σ_0 are obtained, and m and σ_0 are determined by solving Eq. (9) and Eq. (10) as simultaneous equations.

2.4 Results and discussion

In Table 3, the estimators of Weibull parameters are shown.

Table 3 Estimated Weibull parameters

Sample	$\hat{m}$	$\hat{\sigma}_0$
AL1	26.0	211
AL2	22.3	372
AL3	24.2	258
AL4	24.8	262

The experimental Weibull parameters in Table 1 and the estimates of Weibull parameters in Table 3 have a good agreement in the order of magnitude. Therefore, it is shown that this theory is valid. By this theory, fracture strength can be estimated only from the parameters of the initial distribution of flaw size and the number of links.

3. Conclusion

In this paper, the relation between the distribution of the largest flaw size and the distribution of fracture origin size is discussed. By this relation, Weibull parameters are estimated from parameters of the initial distribution of flaw size and the number of links. This method is useful to reliability analysis of brittle materials much easier.

References

[1] Wakabayashi C, Matsuo Y, Yasuda K and Shiota T 2008 *JAME*, **A74** 162
[2] Wakabayashi C, Yasuda K and Shiota T *J.Ceram.Soc. Japan,* **117** 162
[3] Wakabayashi C, Yasuda K and Shiota T 2009 *J. Solid Mechanics and Mat. Eng.* **3** 2
[4] Masuo Y, Kitakami K and Kimura S 1989 *JSME*, **A55** 1
[5] Zhang Y, Inoue M, Uchida N and Uematsu K 1999 *J. Mater. Res.***14** 3370
[6] Abe H, Naito M, Hotta T, Shinohara N and Uematsu K 2003 *J. Am, Ceram. Soc.* **86** 1019

Journal of Physics: Conference Series **191** (2009) 012007 doi:10.1088/1742-6596/191/1/012007

Evaluation of nanoimprint lithography as a fabrication method of distributed feedback laser diodes

M Yanagisawa[1,3], Y Tsuji[1], H Yoshinaga[1], K Hiratsuka[1] and J Taniguchi[2]

[1]Transmission Devices R&D Laboratories, Sumitomo Electric Industries, LTD., 1, Taya-cho, Sakae-ku, Yokohama, 244-8588 Kanagawa, Japan
[2]Faculty of Industrial Science and Technology, Tokyo University of Science, 2641 Yamazaki, Noda, 278-8510 Chiba, Japan

myanagsw@sei.co.jp

Abstract. We have succeeded in employing nanoimprint lithography (NIL) to form the diffraction gratings of distributed feedback laser diodes (DFB LDs) used in optical communication. Uniform gratings and phase-shifted gratings with periods of 232 nm have been formed by using a reverse NIL with a step-and-repeat imprint tool. Line edge roughness has been sufficiently low with the fabricated gratings. DFB LDs fabricated by NIL have indicated comparable characteristics with LDs fabricated by electron beam lithography. We have also demonstrated that phase-shifted DFB LDs show better uniformity in characteristics than uniform-grating DFB LDs. The results of this study indicate that NIL has high potential for the fabrication of DFB LDs.

1. Introduction

Worldwide communication traffic has risen rapidly because of the growth of the Internet, mobile telecommunication, video-on-demand services, and other information-communication facilities. This growth has led to demand for faster and denser communication infrastructures including optical communication networks. Distributed feedback laser diodes (DFB LDs) have the advantages of high selectivity and stability of wavelength; thus, they are widely used as optical sources in networks. They have mainly been used in long-haul and high-speed network nodes, but they are now increasingly required in metro and end-user fields because of increasing traffic. Thus, inexpensive DFB LDs are becoming increasingly necessary.

DFB LDs with uniform-period gratings have been widely used for optical networks. The characteristics of a DFB LD with uniform gratings depend on the grating phase at the cleaved facet [1]. This variation of characteristics with the facet phase is a serious issue from the viewpoints of productivity and usability. One of the most effective ways of reducing the dependence on the facet phase is to convert uniform gratings into phase-shifted gratings [2]. In general, there are various methods for fabricating diffraction gratings, for example, interference exposure, electron beam lithography (EBL), and optical projection exposure. Interference exposure cannot feasibly be used for fabricating phase-shifted gratings, because it exclusively generates exposure patterns with a uniform bright-and-dark period. Although EBL has sufficient resolution to be used for phase-shifted gratings,

[3] Author to whom any correspondence should be addressed.

exceedingly expensive EBL systems are necessary for mass production with sufficient throughput. For the optical projection method, a forefront stepper having sufficient resolution is also expensive, and the cost is too high for fabricating DFB LDs, of which production volume is relatively small, compared to that of such semiconductor devices as LSIs.

Nanoimprint lithography (NIL) has been studied by many organizations since the middle of the 1990s. Chou et al. showed that sub-10-nm features could be formed by imprinting, which triggered the start of NIL technology [3]. A novel method of NIL using a UV-curable polymer was introduced by Haisma et al [4]. Bailey et al. demonstrated a method of step-and-repeat imprinting named SFIL [5].

We started investigating the use of NIL for fabricating diffraction gratings several years ago, because we considered the new technology as a possible solution to the above issues because of its high resolution, throughput, and low cost. In this study, we have used NIL to form phase-shifted gratings of DFB LDs, and we have evaluated its potential as a process for fabricating DFB LDs. We have paid attention to the two technical issues in fabricating DFB LDs: uniformity of residual layer thickness, and mechanical damage in the epitaxial layer caused by imprinting pressure. In this report, we demonstrate the evaluation results of the above issues, and the characteristics of fabricated DFB LDs.

2. Experimental Procedure

Commercially available compound semiconductor substrates, such as GaAs, InP, and GaN, have large undulations compared with Si substrates used for LSIs, of which the total thickness variation typically exceeds 5 μm. When applying NIL to such undulating substrates, it is possible that the mold will come in contact with a limited portion of the substrate, in which case severe nonuniformity of the residual layer thickness in the imprinted area will lead to large variations of the figures in the transferred patterns. Thus, we have formed diffraction gratings on a grating layer by using a reverse imprint based on SFIL/R, developed by Molecular Imprints, in order to suppress the variation of residual layer thickness resulting from the undulation of substrates [6]. We have aimed to improve the uniformity of the figures of grating corrugations by using the SFIL/R process.

Schematic structures of the LD and its epitaxial layers are shown in figures 1(a) and 1(b), respectively. We have prepared an InP substrate with epitaxial layers including an active layer, a grating layer, and a lower cladding layer grown by metalorganic vapor phase epitaxy (MOVPE). The fabrication procedure is shown in figure 2. First, a 50-nm-thick SiON film is deposited on the substrate by plasma-enhanced chemical vapor deposition. Next, a primer material is spin-coated in order to increase adhesion between UV-curable resin and the SiON film. Then, UV-NIL with a quartz glass mold is conducted to form patterns of diffraction gratings on the primer layer [figures 2(a) and 2(b)]. In this study, imprinting was performed in 16 fields on a substrate with a step-and-repeat

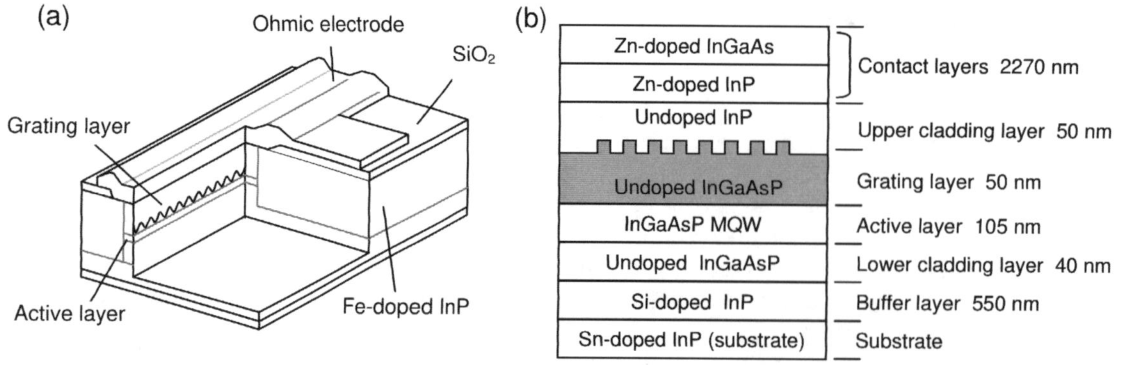

Figure 1. Schematic structure of the DFB LD (a) and its epitaxial layers (b).

equipment. After that, Si-containing resin is spin-coated to cover the grating corrugations [figure 2(c)]. Subsequently, the Si-containing layer is etched by reactive ion etching (RIE) until the tops of the corrugations are revealed [figure 2(d)]. After that, the revealed layer is selectively etched by RIE until the SiON masks are revealed [figure 2(e)], and the formed patterns are used as masks for the subsequent etching, which transfers the grating patterns to the SiON masks [figure 2(f)]. Subsequently, the resin layers are removed by O_2 plasma etching. After that, we use inductively coupled plasma RIE (ICP-RIE) with CH_4/H_2 gas to etch the substrate [figure 2(g)]. Finally, the SiON masks are removed by a wet chemical process using HF solution. After the formation of diffraction gratings, an upper cladding layer and contact layers are formed on the grating layer by MOVPE. The contact layers consist of InP and InGaAs layers [figure 1(b)]. Then, stripe patterns of SiO_2 are formed on the contact layers to define the cavities of the LDs. The stripe patterns are used as masks for subsequent crystal etching by ICP-RIE with CH_4/H_2 gas. After that, Fe-doped InP is selectively grown as an insulating layer by MOVPE. Subsequently, a SiO_2 film is deposited as a passivation layer, in which contact holes are formed by selective etching by RIE. Finally, metal electrodes are formed by high-vacuum evaporation and the lift-off method.

We have formed diffraction gratings with periods from 200 to 260 nm by using molds with corresponding periods. In this study, we used a mold with a period of 232 nm because we aimed to fabricate DFB LDs with a wavelength of 1.49 μm.

As described above, we used the S-FIL/R process so that the variation of the residual thickness has been minimized. However, the effect of the planarization of the S-FIL/R process is not sufficient for fabricating LDs in high yield, and more progress in uniformity of the residual thickness is required. Thus, we have attempted to reduce the field size of the mold.

In order to evaluate the mechanical damage, we have investigated deterioration of photoluminescence (PL) intensities of the epitaxial layers after imprinting.

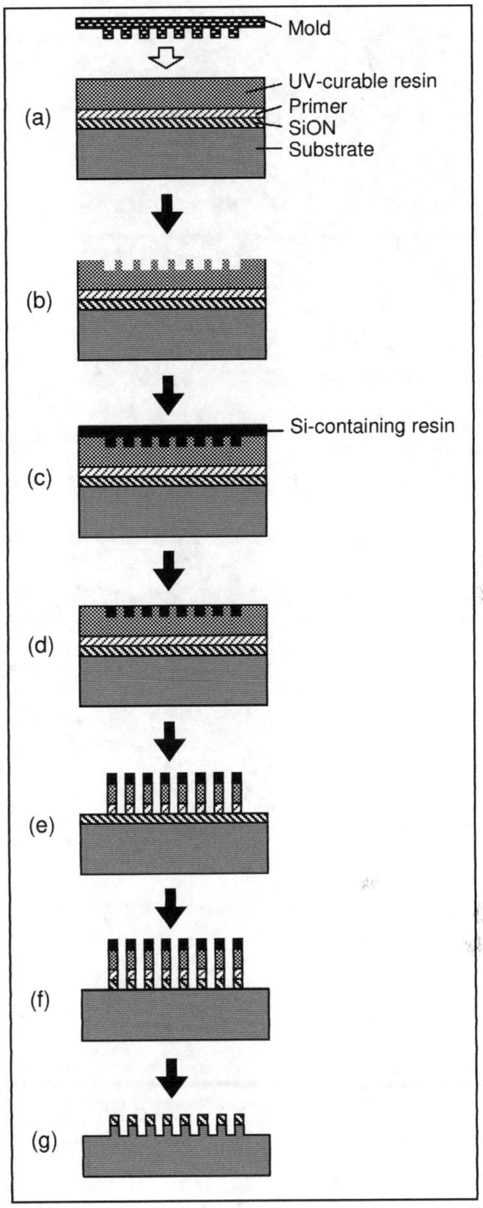

Figure 2. Fabrication process of the gratings.

Journal of Physics: Conference Series **191** (2009) 012007 doi:10.1088/1742-6596/191/1/012007

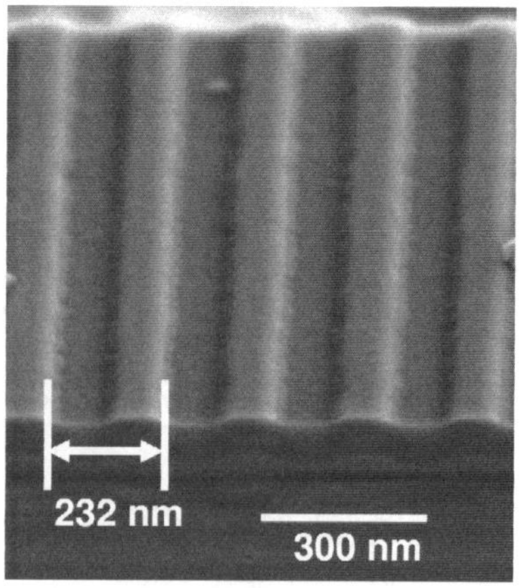

Figure 3. SEM image of the diffraction gratings.

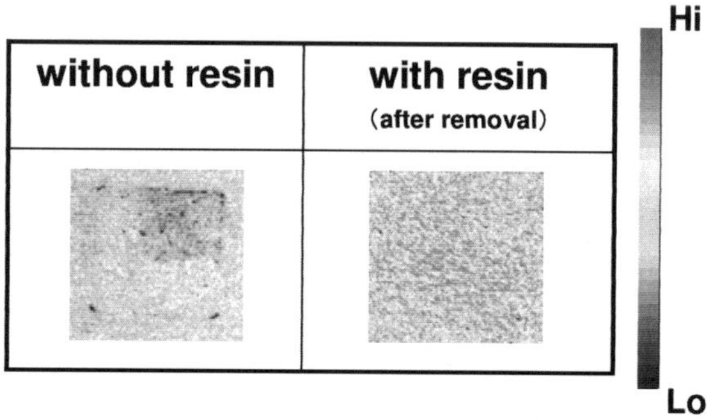

Figure 5. PL intensities of the epitaxial layers after imprinting. The field size of imprinting is 10 mm x 10 mm.

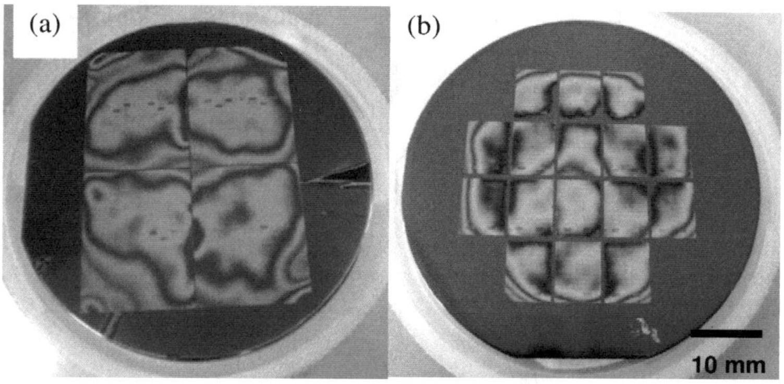

Field size: 14 mm x 18 mm 7 mm x 9 mm

Figure 4. Bird's-eye view of imprinted substrates.

3. Results and Discussion

A scanning electron microscope image of the fabricated diffraction gratings is shown in figure 3. It demonstrates that the line edge roughness (LER) of the gratings is markedly low, indicating that the LER of the grating corrugations in the master mold is sufficiently suppressed and that the master patterns are precisely transferred to the substrate by imprinting. The depth of the diffraction gratings was measured to be approximately 20 nm by an atomic force microscope.

Figures 4(a) and 4(b) are bird's-eye photographs of imprinted substrates. The field sizes in figure 4(a) and figure 4(b) are 14 mm x 18 mm and 7 mm x 9 mm, respectively. They demonstrate that reducing the field size leads to reducing the variation of the residual thickness across the substrate. It is because reducing the field size of a mold corresponds to reducing the variation of the gap between a mold and a substrate within an imprinted field.

We have used a photoluminescence method in order to evaluate the deterioration of crystals. We prepared two indium-phosphide substrates with epitaxial layers. We compared photoluminescence intensities between the two samples: imprinted without resin, and imprinted with resin on it. The results are shown in figure 5. The sample without resin shows evident deterioration of PL intensities. The degradation of intensity is distributed in the edge of the imprinted area. This means that the pressure of imprinting concentrates near the edge of the mold. On the other hand, no evident

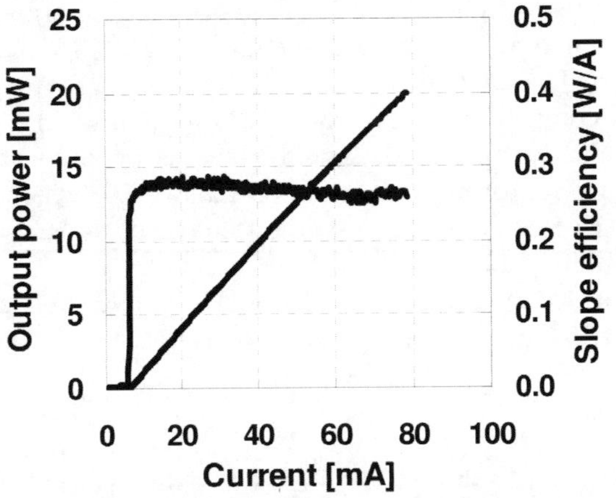

Figure 6. Supplied current vs optical output and slope efficiency.

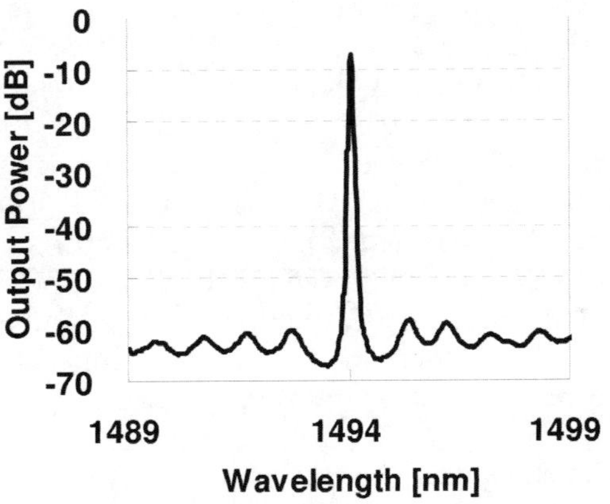

Figure 7. Oscillation spectrum of a phase-shifted LD.

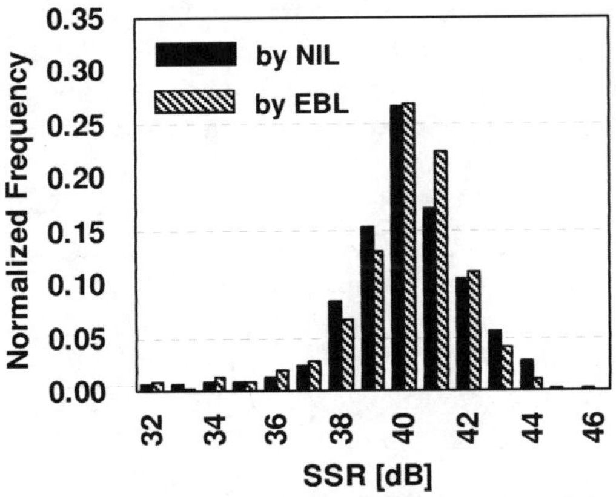

Figure 8. Histograms of SSR for phase-shifted LDs fabricated by NIL and by EBL. The standard deviations of SSR for LDs fabricated by NIL and by EBL are almost the same, 2.0 and 1.8, respectively.

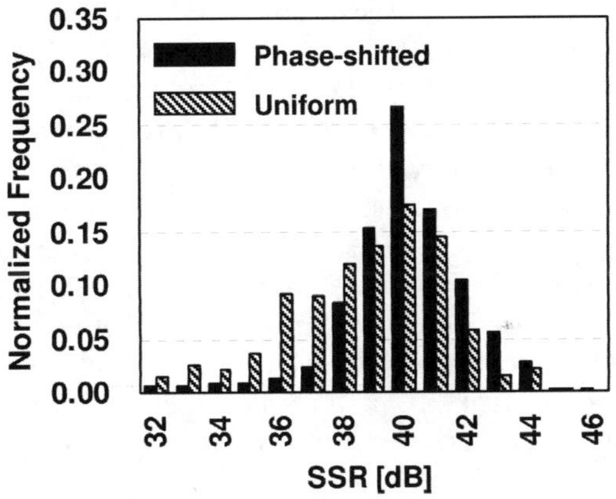

Figure 9. Histograms of SSR for phase-shifted LDs and uniform-grating LDs fabricated by NIL. The standard deviations of SSR for phase-shifted LDs and uniform-grating LDs are 2.0 and 2.5, respectively.

deterioration is found in the sample with resin. These result indicate that the resin functions as a cushion to prevent severe damage in epitaxial layers by imprinting pressure.

Figure 6 shows the dependence of the optical output and slope efficiency on the supplied current for a typical phase-shifted DFB LD fabricated in this study. The threshold current and slope efficiency at room temperature were measured to be 8 mA and 0.28 W/A, respectively, which are comparable to those of typical uniform-grating DFB LDs fabricated simultaneously on the same substrate.

Figure 7 shows the oscillation spectrum of a phase-shifted DFB LD. The resolution of the wavelength is 0.02 nm. This demonstrates that the peak wavelength corresponds to the Bragg wavelength at the center of the stopband, indicating that the phase-shifted gratings function properly.

We have compared the side-mode suppression ratio (SSR) of phase-shifted LDs fabricated by NIL with those fabricated by EBL. SSR is one of the parameters indicating the stability of the single-mode emission of DFB LDs. Histograms of SSR for both types of LD are shown in figure 8, which demonstrate that they show comparable variations in SSR.

The histogram of SSR for phase-shifted LDs is compared with that for uniform-grating LDs in figure 9. The two types of LDs were sampled from the same substrate, and the number of samples is 305 for each type of LD. The standard deviation of SSR for the phase-shifted LDs is 2.0, which is smaller than that for the uniform-grating LDs, 2.5. This demonstrates that the stability of the single-mode emission is improved by the effect of phase-shifted gratings.

As described above, we have successfully fabricated phase-shifted DFB LDs by NIL that are comparable to those fabricated by the conventional EBL process, and we have demonstrated that the uniformity of SSR is improved compared with that for uniform-grating LDs.

4. Conclusions

We have successfully demonstrated the fabrication of phase-shifted DFB LDs by NIL, which have comparable characteristics to those fabricated by EBL. We have also verified the feasibility of fabricating phase-shifted LDs by demonstrating that the uniformity of LD characteristics is improved compared with those of uniform-grating LDs.

NIL is expected to be used as a fabrication process for many applications. However, there are still some difficulties with its use as a mass-production process, for example, defects and low throughput in patterned media, defects and poor alignment accuracy in semiconductor lithography, and the necessary of increasing field size and throughput in displays. Although those difficulties may be common to the fabrication of DFB LDs, they would not be insurmountable problems. Even if defects in the imprinted pattern influence the yield of LDs, inferior chips could be easily rejected because each dye is as small as approximately 300 μm. There is no need for a larger field size than a 3-in.-diameter circle, because compound semiconductor substrates used for LDs are wafers of 3-in. diameter or smaller. Regarding alignment accuracy, an error of up to approximately 5 μm is acceptable. Also, even if throughput is limited to less than one wafer per hour, NIL would still have a higher throughput than EBL.

As described above, NIL is an effective and promising process for fabricating phase-shifted DFB LDs and is expected to have the advantage of mass-production capability in the near future. We conclude that NIL has high potential for fabricating DFB LDs, and we expect that NIL will be used for fabricating various optical devices consisting of nanostructures.

References

[1] Matsuoka T, Nagai H, Noguchi Y, Suzuki Y and Kawaguchi Y 1984 *Japan. J. Appl. Phys.* **23** L138

[2] Kaden C, Griesinger U, Schweitzer H, Pilkuhn M H and Stath N 1992 *J. Vac. Sci. Technol.* B **10** 2970

[3] Chou S Y, Krauss P R and Renstrom P J 1995 *Appl. Phys. Lett.* **67** 3114

[4] Haisma J, Verheijen M, Heuvel K and Berg J 1996 *J. Vac. Sci. Technol.* B **14** 4124

[5] Bailey T, Choi B J, Colburn M, Meissl M, Shaya S, Ekerdt J G, Sreenivasan S V and Willson C G 2000 *J. Vac. Sci. Technol.* B **18** 3572

[6] Miller M, Doyle G, Stacey N, Xu F, Sreenivasan S V, Watts M and LaBrake D L 2005 *Proc. SPIE* 5751 994

Journal of Physics: Conference Series **191** (2009) 012008 doi:10.1088/1742-6596/191/1/012008

A Molecular Dynamics Study on the Effects of Nanostructural Clearances on Thermal Resistance at a Liquid Water-Solid Interface

M Shibahara[1], **K Takeuchi**[1]
[1]Osaka University, 2-1 Yamadaoka, Suita, Osaka 565-0871, Japan

E-mail: siba@mech.eng.osaka-u.ac.jp

Abstract. The classical molecular dynamics simulation was conducted in order to clarify the effects of surface structural clearances at the nanometer scale on thermal resistance at a liquid water-solid interface as well as static and dynamic behaviours of fluid molecules in the vicinity of the surface. A liquid molecular region confined between two solid walls, of which the interparticle potential was Lennard-Jones type or SPC/E type, was employed as a calculation system. The thermal resistance between the liquid molecular region and the solid walls with nanostructures was calculated by the heat flux and the temperature discontinuity obtained in the molecular dynamics simulations. With changing the surface structural clearances and the potential energy functions the thermal resistance at the interface once decreased and got to the minimum value when the structural clearances were changed from 0 to several nm.

1. Introduction

Due to ongoing progress in material surface treatments and in nano/micro scale processing, it is feasible to fabricate various structures at nano- or micrometer scale on a solid surface. On the other hand, the molecular dynamics method has been applied to understand static structures and transport phenomena at the molecular scale at a liquid-solid interface, and a substantial thermal resistance was found at the liquid-solid interface, as well as between solid-solid interfaces when observed at the molecular scale [1-7]. The authors have previously reported that there was some energy transfer difference at a liquid-solid interface depending on surface structures and structural clearances at the nanometer scale when we observe the energy transfer at the interface [8-11]. Effects of surface structures and structural clearances at the nanometer scale on the interfacial thermal resistance between liquid and solid were also discussed from the molecular dynamics point of view [12-13]. In the present study a liquid molecular region confined between the solid walls, of which the intermolecular potential was Lennard-Jones type or SPC/E type, was employed as a calculation system. The temperature of the middle layers of the upper and the lower solid walls were controlled by the Langevin method [14,15] in order to make a temperature gradient in the calculation system. Effects of surface structures and surface structural clearances on the thermal resistance between liquid and solid as well as static structures and dynamic behaviours of fluid molecules were investigated directly by the classical molecular dynamics method.

2. Numerical methods

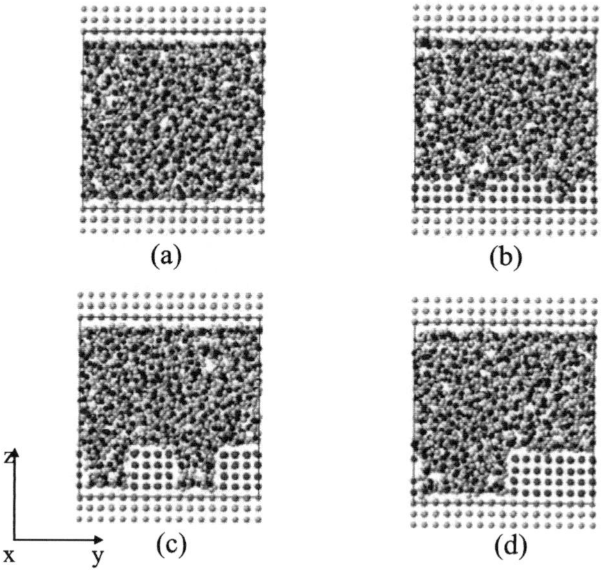

Fig.1 Calculation models of surface structures at the nanometer scale and snapshots in liquid molecular region.

Figure 1 shows one of calculation models employed in the present study and it shows surface structures at the nanometer scale and snapshots in liquid molecular region too. A liquid molecular region of which the size was 2.8×2.8×2.8 nm^3 or 5.6×5.6×5.7 nm^3 was confined with upper and lower solid walls as a unit cell. Periodic boundary conditions were employed in x and y directions. Upper and lower solid walls consisted of three solid layers where the outermost layers were fixed and the temperatures of the middle layers were controlled at constant by the Langevin method. We assumed that there were various nanostructures of 0.7 nm height attached to the lower solid wall and the structural clearances were changed from 0.00 to 2.81 nm. As shown Fig. 1 the nanostructures attached to the surface have the periodic groove structures in y direction. When the initial surface structural clearance was less than 0.4 nm in case of surface (b) in Fig. 1, the surface nanostructures were so small that they were broken down into random structures. Nanostructures in the unit cell consisted of the same number of atoms. Density in the liquid molecular region was kept constant and the effects of liquid density on the thermal resistance were already discussed in the previous study [12]. The temperature gradients were formed in the liquid molecular region by controlling the temperatures of the upper wall T_l and the lower walls T_h, respectively. The difference between T_l and T_h was adjusted to about 50K in the present study. The numerical uncertainty of the temperature and the temperature gradient obtained was directly dependent on the total calculation time and it was within 5 % in the present calculation time. The 12-6 Lennard-Jones potential energy functions and SPC/E potential functions with cutoff distances were employed in the present study so as to understand the energy transfer phenomena qualitatively. The 12-6 Lennard-Jones potential parameters of liquid molecules correspond to those of water molecules and potential parameters of the wall atoms correspond to those of iron atoms. We employed the 12-6 Lennard-Jones potential functions and the SPC/E potential functions with cut off distances for the rigid water molecules [16,17] so as to investigate the effects of the potential functions between liquid molecules on the calculated thermal resistance. Temperature gradients and energy transfer at upper and lower walls were calculated numerically after a relaxation calculation of 1.0 ns. Newton's equations were integrated numerically by the Verlet method with a time interval of 5 fs for Lennard-Jones water models and 0.1 fs for SPC/E water models,

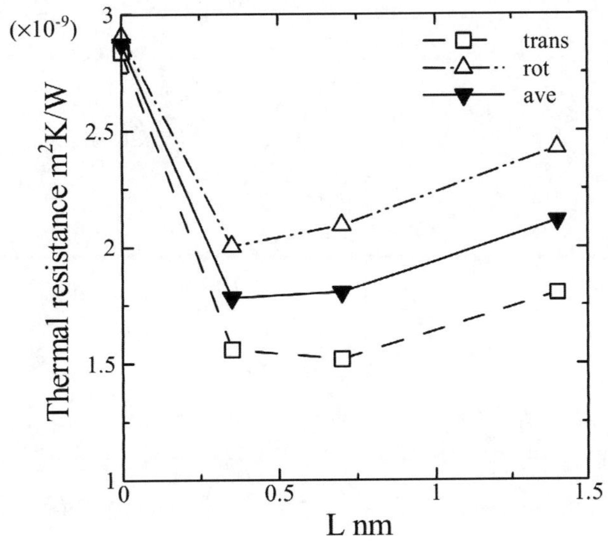

Fig.2 Effects of nanostructural clearances on thermal resistances calculated by rotational, translational and the average temperatures.

respectively. The thermal resistance R_t between a solid wall and a liquid region was calculated by the following equations as,

$$R_t = \frac{\Delta T}{Q}, \quad (1)$$

where ΔT is the temperature discontinuity between the solid wall and the liquid region and Q is the energy flux that was added or subtracted by the Langevin method per unit area (divided by the unit cell size in x-y plane) so as to control the solid wall temperature at constant. We can obtain a certain temperature gradient in the liquid region as well as in the solid walls during a molecular dynamics simulation. By using the temperature gradient obtained by the molecular dynamics simulation we can calculate the temperature discontinuity ΔT between the solid wall and the liquid region as a result. The energy flux in the calculation domain Q can be calculated by the energy input at the lower wall (heating wall). We checked that the energy output at the upper wall (cooling wall) exactly coincided with the energy input at the lower wall. The similar calculation method of the thermal resistance was employed in the molecular dynamics simulations and reported by Maruyama et al. [3,4] and the similar calculation method of the temperature gradients was also applied in the molecular dynamics simulation and reported by Ohara et al. [5,6].

3. Calculation results

3.1 Effects of Nanostructural Clearances on Thermal Resistance Depending on Degrees of Freedom of Water Molecules

Figure 2 shows effects of the nanostructural clearances on the thermal resistances calculated by rotational temperature, translational temperature and the average temperature when the intermolecular potential functions are the SPC/E potential with cut off distance r_c=1.3nm. The rotational temperature T^{rot}, the translational temperature T^{trans} and the average temperature T^{ave} can be calculated by the following equations,

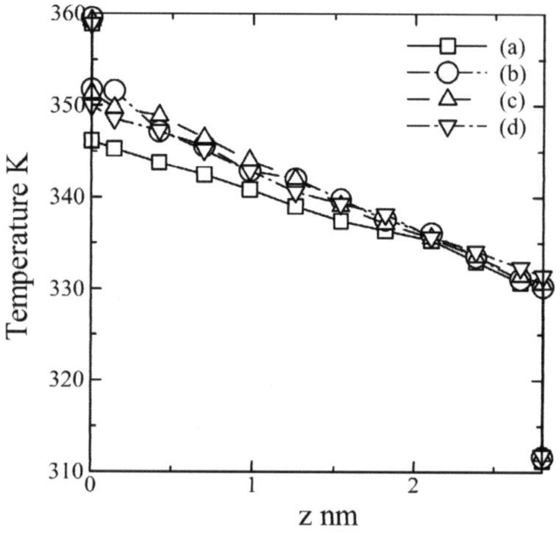

(1) Translational temperature

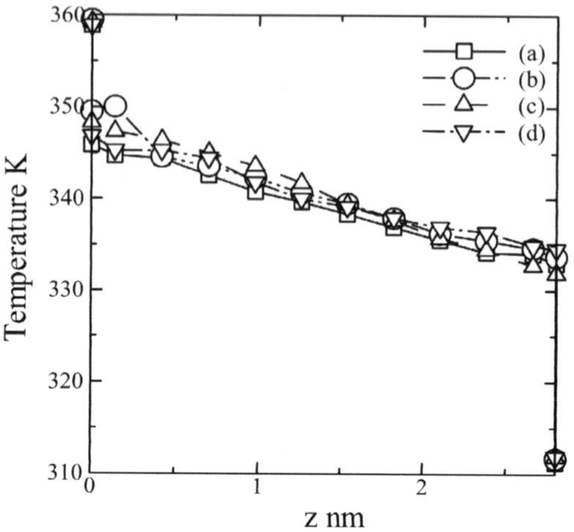

(2) Rotational temperature

Fig.3 Effects of the nanostructural clearances on the gradients of translational temperature and rotational temperature.

$$\frac{v_r k}{2} T^{rot} = \frac{1}{N} \sum_i \frac{1}{2} \mathbf{I} \boldsymbol{\omega}_i^2, \qquad (2)$$

$$\frac{v_t k}{2} T^{trans} = \frac{1}{N} \sum_i \frac{1}{2} m \mathbf{v}_i^2, \qquad (3)$$

$$\frac{(v_r + v_t) k}{2} T^{ave} = \frac{1}{N} \left(\sum_i \frac{1}{2} \mathbf{I} \boldsymbol{\omega}_i^2 + \sum_i \frac{1}{2} m \mathbf{v}_i^2 \right), \qquad (4)$$

where v_r, v_t, N, m, $\mathbf{v}$, $\mathbf{I}$ and $\boldsymbol{\omega}$ are rotational degrees of freedom, translational degrees of freedom, number of water molecules, mass of a water molecule, translational velocity, moment of inertia and angular velocity, respectively. The nanostructural clearances L are changed as 0.00, 0.35, 0.70 and 1.40 nm. Figure 3 shows the gradients of translational temperature and rotational temperature depending on the surface nanostructural clearances. All thermal resistances calculated by the translational temperature, the rotational temperature and the average temperature of water molecules once decreased and got to the minimum value and increased again with the increase of the nanostructural clearances and the order of the calculated thermal resistances was in the range from 1.5×10^{-9} to 3×10^{-9} m^2K/W. The thermal resistance calculated by the rotational temperature was larger than that by the translational temperature under the condition of the same nanostructural clearance. In the case of L = 0.00 nm, in other words a flat surface, the thermal resistance calculated by the translational temperature coincided with that by the rotational temperature. Hence, the thermal resistance differences depending on degrees of freedom of water molecules, the translational and rotational temperature, came from the interaction between the nanostructural clearances and the water molecular dynamic motions. The rotational motions of water molecules in the nanostructural clearances were somewhat restricted although the effective surface area in the molecular scale was enlarged by the nanostructures.

3.2 Effects of Nanostructural Clearances on Thermal Resistance Depending on Potential Energy Functions between Water Molecules

Figure 4 shows the effects of the nanostructural clearances and the potential energy functions between water molecules on the thermal resistances at a liquid water-solid interface. Potential energy functions

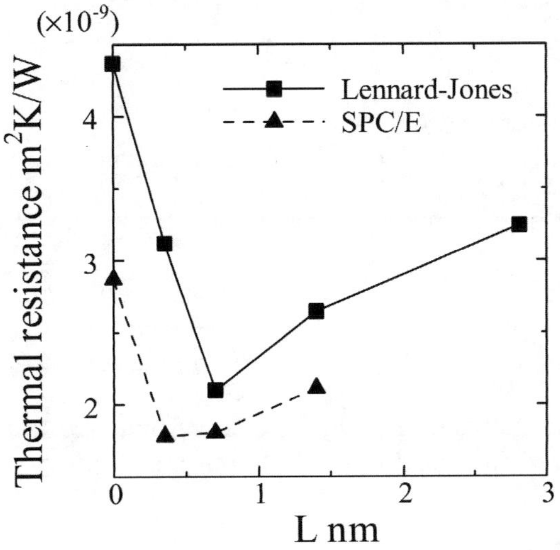

Fig.4 Effects of the nanostructural clearances and the potential energy functions between water molecules on the thermal resistances at a Liquid water-solid interface.

between water molecules employed in the present study were 12-6 Lennard-Jones potential function and SPC/E potential function with the cut off distance $r_c = 1.3$ nm. In the case of SPC/E potential the thermal resistances were calculated by the average temperature of translational and rotational temperature of water molecules. Regardless of the potential energy functions between water molecules and the nanostructural clearances the order of the thermal resistances calculated was in the range from 1.5×10^{-9} to $4.5\times10^{-9}m^2$ K/W. The thermal resistances calculated once decreased and got to the minimum value and increased again in both cases of the potential energy functions. The thermal resistances with nanostructures were lower than those of flat surface in the cases of both potential functions. The thermal resistance at a liquid water-solid interface has much relationship with the surface area enlargement and the mobility change of water molecules depending on the nanostructural clearances.

4. Conclusions

The classical molecular dynamics simulation was conducted in order to clarify effects of surface structural clearances at the nanometer scale and potential energy functions on a thermal resistance at a liquid water-solid interface. A liquid molecular region confined between two solid walls, of which the intermolecular potential was Lennard-Jones type or SPC/E type, was employed as a calculation system. A thermal resistance at a liquid water-solid interface with nanostructures was calculated by the heat flux and the temperature discontinuity obtained by the molecular dynamics simulation. The surface structures and the structural clearances were changed as calculation parameters. The main findings were as follows:

(1) There was a certain difference between the thermal resistance calculated by the translational temperature and that by the rotational temperature depending on the interfacial structural clearances when the SPC/E potential functions were employed as the intermolecular potential functions between water molecules.

(2) The thermal resistance at the liquid water-solid interface passed through a minimum value when the interfacial structural clearances were changed at the nanometer scale.

Acknowledgement

This work was partially supported by the Ministry of Education, Science, Sports and Culture (MEXT), Japan, "Grant-in-Aid for Young Scientists (A) No. 19686017".

References

[1] Pollack G L 1969 *Reviews of Modern Physics* **41** 48
[2] Xue L, Keblinski P, Phillpot S R, Choi S U -S and Eastman J A 2004 *International Journal of Heat and Mass Transfer* **47** 4277
[3] Maruyama S, Kimura T, Ho C S and Yasui K 1998 *Proceedings of JSME Thermal Engineering Conference* 309
[4] Maruyama S and Kimura T 1999 *Thermal Science & Engineering* **7** 63
[5] Ohara T and Yatsunami T 2003 *Microscale Thermophysical Engineering* **7** 1
[6] Torii D, Ohara T and Ishida K, 2006 *Proc. of 13th International Heat Transfer Conference* CD-ROM
[7] Nagayama G and Cheng P 2004 *International Journal of Heat and Mass Transfer* **47** 501
[8] Shibahara M, Kunugi T, Muko K and Katsuki M 2006 *JSME International Journal Series B* **49**-2 506
[9] Shibahara M, Kunugi T, Muko K and Katsuki M 2006 *Microscale Thermophysical Engineering* **10**-3 197
[10] Shibahara M, Kunugi T, Muko K and Katsuki M 2005 *Transaction of the Japan Society of Mechanical Engineering Series B* **71**-708 2108
[11] Shibahara M, Kunugi T, Muko K and Katsuki M 2005 *Transaction of the Japan Society of Mechanical Engineering, Series B* **71**-708 2112
[12] Shibahara M, Inoue K and Kobayashi K 2008 *Proceedings of ASME Micro/Nanoscale Heat Transfer International Conference* CD-ROM
[13] Shibahara M and Inoue K 2008 *Transaction of the Japan Society of Mechanical Engineering* Series B **74**-737 172
[14] Tully J C 1980 *Journal of Chemical Physics* **73**-4 1975
[15] Maruyama S 2000 *Advances in Numerical Heat Transfer* **2** 189
[16] Kotsalis E M, Walther J H and Koumoutsakos P 2004 *International Journal of Multiphase Flow* **30** 995
[17] Walther J H, Jaffe R, Halicioglu T and Koumoutsakos R 2001 *J. Phys. Chem.* **105**-41 9980

Journal of Physics: Conference Series **191** (2009) 012009 doi:10.1088/1742-6596/191/1/012009

Application of FIB Technique to Introduction of a Notch into a Carbon Fiber for Direct Measurement of Fracture Toughness

S Ogihara[1], Y Imafuku[2], R Yamamoto[2] and Y Kogo[3]

[1]ogihara@rs.noda.tus.ac.jp

[1] Department of Mechanical Engineering, Tokyo University of Science,

2641 Yamazaki, Noda, 278-8510 Chiba, Japan

[2] Graduate Student, Tokyo University of Science

[3] Department of Materials Science, Tokyo University of Science

Abstract. For the direct measurement of the fracture toughness of the carbon fiber, a new technique was proposed and examined its applicability. At first, machining condition of the notch was examined. The notch was introduced using focused ion beam (FIB). The ion beam can be electronically scanned to introduce a sharp notch on the carbon fiber. Notches with various notch width and length were introduced by changing beam and scanning conditions. Tensile tests on notched carbon fibers were carried out following the test method for carbon fiber monofilaments. Fractured specimens were successfully corrected without secondary damage using protection films. SEM observations revealed that a crack propagated from a notch-tip, and notch size was able to be determined successfully. Effect of notch root radius was also examined to investigate the validity of the fracture toughness obtained by this method.

1. Introduction

Fracture toughness of carbon fiber is of great interest as well as its strength, because it is essentially important when fracture behavior of composites was investigated. For example, fracture mode of brittle matrix composites are strongly affected by the fiber fracture toughness as well as its interface toughness [1]. The measurement of fiber fracture toughness is, however, thought to be difficult due to its small diameter. One of the practical ways to estimate fracture toughness is determination of the initial crack size by observing fracture mirror size on a tensile fracture surface of un-notched specimens [2]. Since the mirror size measurement involves a considerable error, estimated fracture toughness is usually scattered significantly. Accuracy of the measured fracture toughness will be significantly improved, if a notch is introduced intentionally by controlling its size and shape. In this study, therefore, a new measurement technique was proposed to estimate fracture toughness of the carbon fiber.

At first, applicability of the focused ion beam (FIB) system was examined as a machining tool of notches. The FIB is regarded as a useful machining tool to prepare samples for transmission electron microscope observations. Recent FIB systems have less than 10 nm beam diameter, achieving micro-machining by scanning the beam electronically. Various sizes and types of notches were tried to be introduced on the carbon fibers. Tensile tests were carried out on the notched carbon fibers to examine the validity of this technique.

2. Experimental procedure

2.1 Specimens Preparation

The carbon fiber tested in this experiment is a PAN based high strength type carbon fiber (IM-600; 6000 filaments) manufactured by *Toho Tenax Co. Ltd, Japan.* A carbon monofilament was extracted from a bundle, and was attached on a holder as shown in Figure 1. An aluminium holder and carbon tapes were used since electrical conductivity was required to introduce a notch by the FIB.

The notch was introduced using the FIB system (JFIB-2300; *JEOL Co. Ltd., Japan*) on a carbon fiber. Ga^+ ion beam was extracted from an ion source and accelerated with 30kV accelerating voltage. Various types of notches were introduced by changing a beam diameter and other machining conditions. Details of the conditions were shown in Table 1.

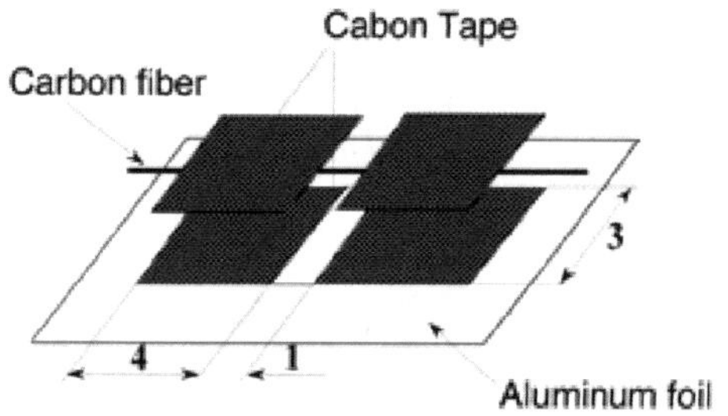

Figure 1. Schematic illustration of facture toughness specimen

Table 1. FIB machining conditions

Acceleration Voltage	30 kV
Extraction Voltage	5.58 kV
Emission Current	1.26 μA
Ion Beam Current	0.6 μA
Filament Current	1.62 A
Filament Voltage	0.58 V
Prove Current	1 pA
Beam Diameter	7 ~ 18 nm

2.2 Fracture Toughness Test

Figure 2 schematically shows a cross section of the carbon fiber on a notched plane. A notch was introduced by the FIB system on one side surface of the carbon fiber with various notch depths (a) as indicated by gray area in Figure2. Notched carbon fibers were tested in tension with crosshead speed of 0.1 mm/min. following the tensile test method of carbon fiber monofilaments. Plastic films were set on both sides of a carbon fiber specimen, and liquid detergent was filled between them to avoid secondary damage of the carbon fiber. After tensile tests, liquid detergent was removed by dipping in water. Fracture surface was observed by SEM to confirm a fracture position and to determine the notch length, a.

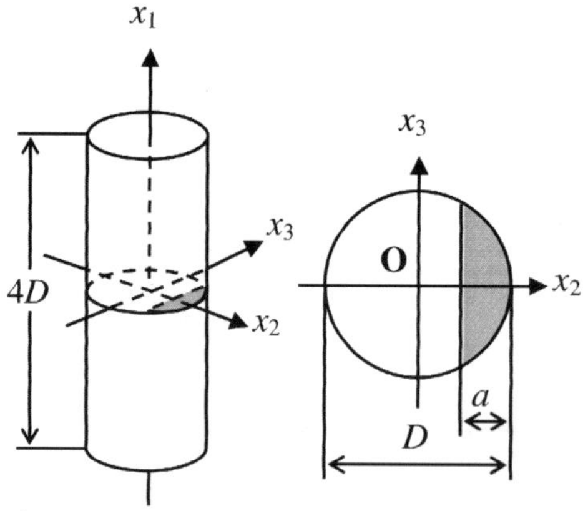

Figure 2. Schematic illustration of notched carbon fiber

3. Finite elements analysis

3.1 Calculating fracture toughness

Energy release rate, G_I, at the center of the crack front was examined by the finite elements method using the virtual crack closure method (VCCM). Figure 2 shows the model of a carbon monofilament which has a straight edge crack from the surface located at its center. The model was analyzed as both an isotropic and an orthotropic materials. In the case of the isotropic material, Young's modulus (E) is set to be 285GPa, Poison's ratio (ν) 0.3. Table 2 shows the material properties used for the analysis in the case of the orthotropic material. In the case of isotropic material, the energy release rate ($G_{\rm I}$) is converted to the stress intensity factor ($K_{\rm I}$) by the following equation

$$G_{\rm I} = \frac{K_{\rm I}^{2}\left(1-\nu^{2}\right)}{E} \tag{1}$$

where assumption of plain strain condition in the neighborhood of the crack front is made. In the case of orthotropic material, following equation is used [3].

$$G_{\rm I} = K_{\rm I}^{2}\sqrt{\frac{a_{11}a_{22}}{2}}\left[\sqrt{\frac{a_{22}}{a_{11}}}+\frac{2a_{12}+a_{66}}{2a_{11}}\right]^{1/2} \tag{2}$$

where

$$a_{11} = \frac{1}{E_{11}},\ a_{22} = \frac{1}{E_{22}},\ a_{66} = \frac{1}{G_{12}},\ a_{12} = a_{21} = -\frac{\nu_{12}}{E_{11}} = -\frac{\nu_{21}}{E_{22}} \tag{3}$$

Table 2. Material properties used for orthotropic analysis

Young's modulus [GPa]	Poisson's Ratio	Shear modulus [GPa]
E_{11} 285	ν_{12} 0.300	G_{12} 20.0
E_{22} 30.0	ν_{23} 0.400	G_{23} 10.7
E_{22} 30.0	ν_{31} 0.0316	G_{31} 20.0

The dimensionless stress intensity factor which is described by

$$f = \frac{K_{\rm I}}{\sigma\sqrt{\pi a}} \tag{4}$$

where σ is the far field stress and a is the crack length. Figure 3 shows the comparison between f calculated by equation (1) (isotropic case) and that in [4]. A good agreement is obtained which implies the accuracy of the present calculation. Figure 4 shows the comparison of f between isotropic and orthotropic cases. The dimensionless stress intensity factor is approximated as

$$f = \frac{K_I}{\sigma\sqrt{\pi a}} = 28.3\left(\frac{a}{D}\right)^{3} - 15.0\left(\frac{a}{D}\right)^{2} + 4.09\left(\frac{a}{D}\right) + 0.773 \tag{5}$$

in the isotropic case and as

$$f = \frac{K_I}{\sigma\sqrt{\pi a}} = 9.19\left(\frac{a}{D}\right)^{3} - 2.87\left(\frac{a}{D}\right)^{2} + 0.859\left(\frac{a}{D}\right) + 0.511 \tag{6}$$

in the orthotropic case. Fracture toughness was calculated by these approximations.

3.2 Fiber strength

Fracture stress of notched fiber σ_0 is described by

$$\sigma_0 = \frac{\sigma_f}{\alpha} \tag{7}$$

where σ_f is the fiber strength and α is stress concentration factor at the notch tip. We calculated the stress concentration factor using the FEM analysis of the notched fiber. We assumed σ_f is 6GPa. The fracture stress σ_0 can be calculated by equation (7). Using the fracture stress σ_0, fracture toughness is calculated by approximations (5) and (6).

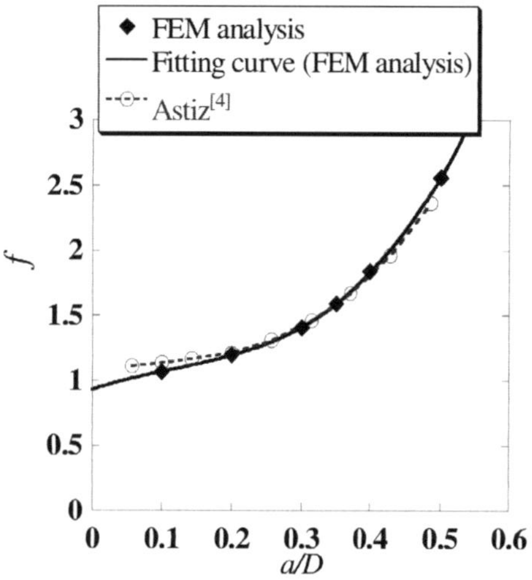

Figure 3. Comparison of the dimensionless stress intensity factor (isotropic)

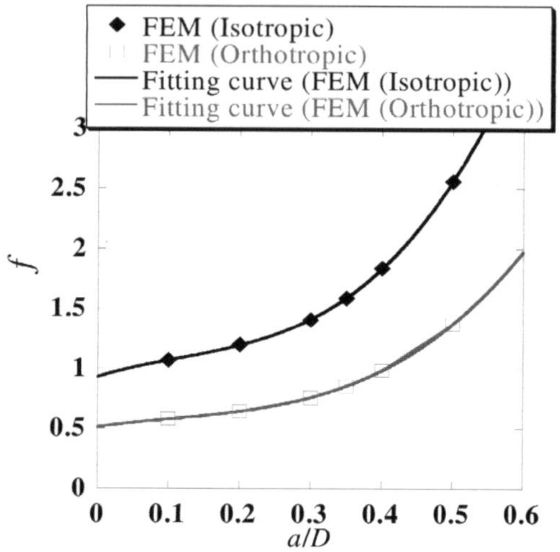

Figure 4. Comparison of the dimensionless stress intensity factors between isotropic and orthotropic cases

4. Results and discussion

4.1 Notch machining

Figure 5 shows typical outlooks of notched carbon fibers introduced with different beam condition. When the beam diameter of 7nm was used, a sharp notch was successfully introduced with notch-tip width of less than 50 nm as shown in (a). On the other hand, 360nm notch-tip width was obtained by changing a beam diameter (18 nm) and scan area as shown in (b). Various kinds of notches were introduced with different notch-width and notch-length.

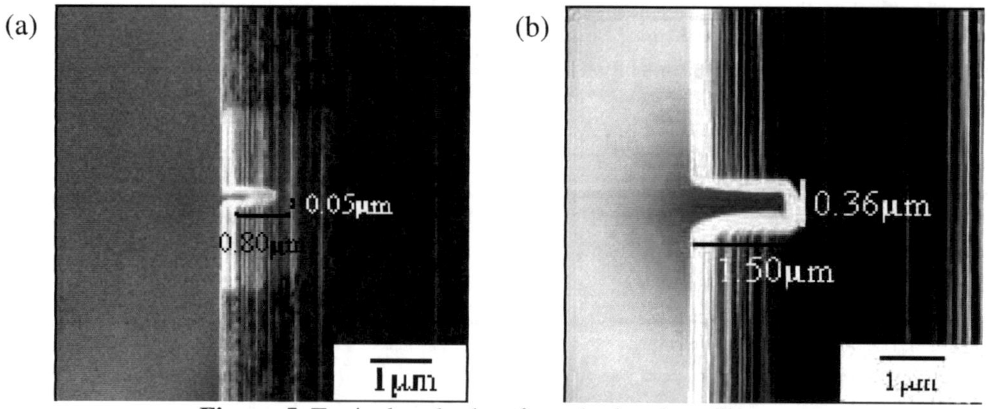

Figure 5. Typical outlooks of notched carbon fibers
(a) introduced by the beam with 7 nm diameter,
(b) by the beam with 18 nm diameter

4.2 Tensile tests

A typical load-displacement (P-δ) curve was shown in Figure 6. The P-δ curve was linear up to the final fracture. In all tests, same behavior was observed. After tensile tests, specimens were successfully corrected without secondary damage and fracture surface could be observed as shown in Figure 7. A flat notch surface was clearly observed as indicated by a circle on the fracture surface. It was also observed that the notch front was straight, indicating machining by the FIB was applicable to the notch introduction. Notch length, a, was measured from the micrograph for every specimen.

On the fracture surface, hackle pattern was clearly observed from the notch tip.

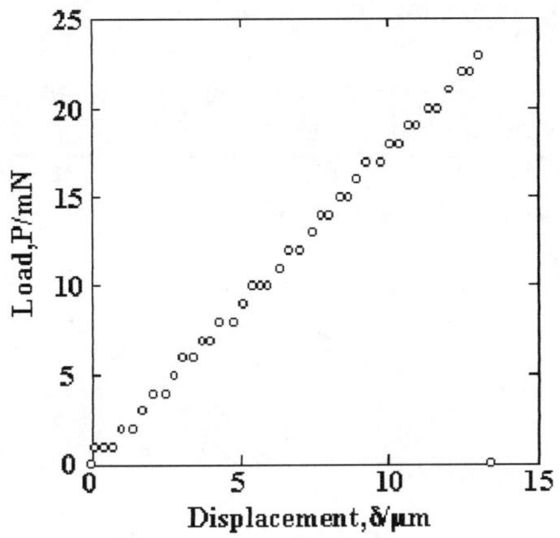

Figure 6. Typical Load-displacement curve of notched carbon fiber in tension test

Figure 7. Typical Fracture surface of notched carbon fiber tested in tension.

This means that the fracture of the notched carbon fiber originated from a notch-tip. It was confirmed with these observations that fracture toughness tests were successfully carried out with the present technique.

Based on the fracture toughness tests with various widths of notches, effect of notch width was examined as shown in Figure 8. Although K_C was somewhat scattered, it showed almost constant value of 1.6 $\mathrm{MPa}\sqrt{\mathrm{m}}$ (isotropic) and 1.0 $\mathrm{MPa}\sqrt{\mathrm{m}}$ (orthotropic) on average, when the notch-tip width was less than 0.2 μm. Since the fracture toughness of the PAN based carbon fibers were reported to be in the rage from 1~2 $\mathrm{MPa}\sqrt{\mathrm{m}}$ [2], present results is expected to give the reasonable value in this range of notch width. On the other hand, K_C linearly increased with notch-tip width, if the notch-tip width was more than 0.2 μm. These results suggest that the fracture toughness tests should be carried out with small enough notch-tip width to which calculated fracture toughness is insensitive.

The fracture toughness was predicted by calculating the stress concentration in a notched carbon fiber. It was found that the predicted fracture toughness was much lower than the experimental value when the strength of the carbon fiber was assumed to be 6 GPa (Figure 8).

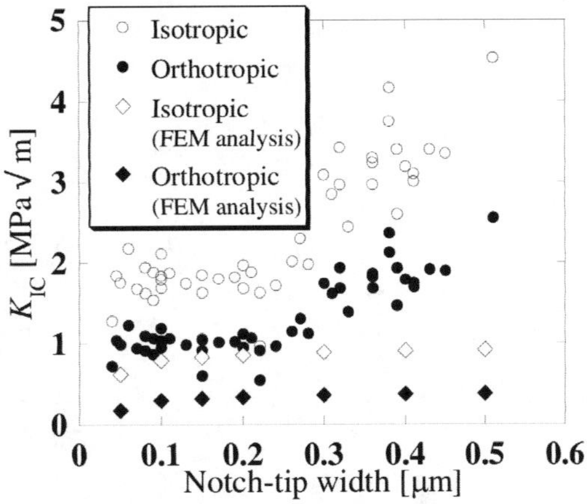

Figure 8. Fracture toughness of a carbon fiber

5. Conclusion

(1) Various types of notches with straight notch front could be successfully introduced on carbon fiber monofilaments using a focused ion beam machining system.
(2) Fracture surface observation revealed that fracture of notched carbon fibers originated from notch tip with hackle pattern, indicating present technique is applicable for fracture toughness tests of the carbon fiber.
(3) Effect of notch tip width was clearly shown that the fracture toughness tests should be carried out with small enough notch tip width to which calculated fracture toughness is insensitive.
(4) Stress intensity factor is calculated by using the VCCM technique as a function of crack length for both isotropic and orthotropic carbon fibers. The fracture toughness of IM600 was estimated as $1.6\ \mathrm{MPa}\sqrt{\mathrm{m}}$ (isotropic), $1.0\,\mathrm{MPa}\sqrt{\mathrm{m}}$ (orthotropic) on average.

6. References

[1] He M., Hutchinson J.W. 1989 *J. Appl. Mech.* **56** 270
[2] Honjo K 2003 *Carbon* **41** 979.
[3] Sih G.C., Paris P.C., Irwin G.R. 1965 *Int. J. Fract.* **1** 189
[4] Astiz M.A. 1986 *Int. J. Fract.* **31** 105

Journal of Physics: Conference Series **191** (2009) 012010 doi:10.1088/1742-6596/191/1/012010

Study of Reactive Ion Etching for Reverse Tone Nanoimprint Process

Y Tsuji, M Yanagisawa, H Yoshinaga, and K Hiratsuka

Fiber-optics Core Devices R&D Department,
Transmission Devices R&D Laboratories, Sumitomo Electric Industries, LTD.
1, Taya-cho, Sakae-ku, Yokohama, Kanagawa, 244-8588 Japan

E-mail : tsuji-yukihiro@sei.co.jp

Abstract: We have used reverse nanoimprint for fabricating diffraction gratings of distributed feedback laser diodes. Generation of residues in the etching process of resin is a serious issue leading to poor line edge roughness of the grating patterns. We have found that the residues are composed of oxide products from Si-containing resin. We have successfully suppressed the generation of the residues by optimizing oxygen partial pressure of reactive ion etching (RIE). We have also succeeded in effectively removing the residues by utilizing sputtering effect of RIE.

1. Introduction

Nanoimprint lithography (NIL) has been studied to fabricate diffraction gratings of distributed feedback laser diodes (DFB LDs) for optical communication network [1]. Because compound semiconductor wafers such as GaAs and InP for device fabrication have a certain undulation, thickness uniformity of imprinted residual layer is affected. Then, reverse tone nanoimprint process, for example SFIL/R^{TM} process proposed by Molecular Imprints, Inc. [2], has been applied in order to improve the uniformity of residual layer thickness. There are two kinds of resin layers stacked on the wafer and the two-step reactive ion etchings (RIE) have been used for pattern transfer. The upper layer is a Si-containing resin, which is used for an etching mask in the second RIE.

In this paper, we demonstrate the optimized etching condition for fabricating diffraction gratings in view of eliminating resin residues.

2. Experimental Procedure

Reverse tone nanoimprint process is shown in figure 1. We prepared 2-in-diameter InP epitaxial wafers including InGaAsP multi-quantum-well (MQW) layer, the function of which is not only an active layer but also a grating layer. Grating patterns with 200-240 nm pitch have been transferred on the wafers by a serial process described in below. An 80 nm-thick film of adhesion layer is spin-coated on the wafer [figure 1(a)]. An 8 mm x 6 mm template is pressed on dispensed UV curing resin at a force of 4 N [figure 1(b) and 1(c)]. After demolding, the spin-coating of a 200 nm-thick Si-containing resin is followed by the etch-back by first RIE to reveal the tops of the imprinted feature, which is selectively etched down to the wafer by subsequent RIE using oxygen [figure 1(d), 1(f), 1(g) and 1(h)]. These grating patterns are used as masks for the etching of the wafer [figure 1(i)].

Journal of Physics: Conference Series **191** (2009) 012010 doi:10.1088/1742-6596/191/1/012010

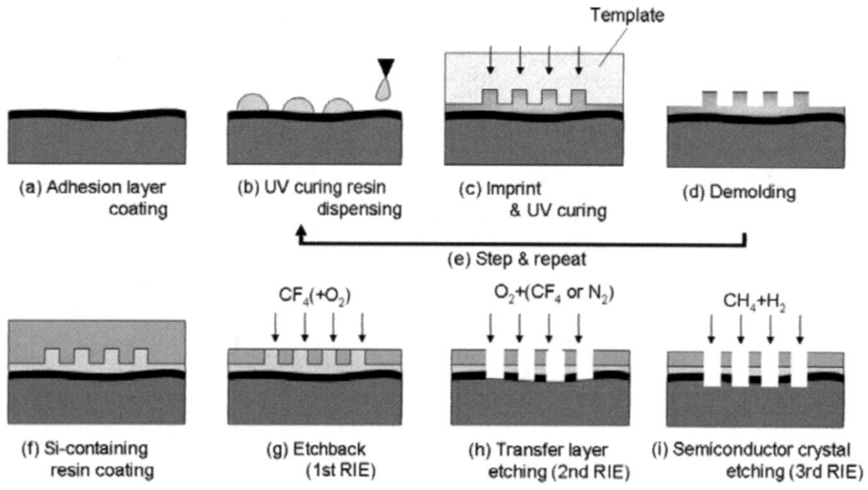

Figure 1. Reverse tone nanoimprint process

The second RIE is carried out using the inductively coupled plasma (ICP) RIE system. The etching rate of the adhesion layer is approximately 200 nm/min and the etching selectivity ratio is more than 30. The etching time is 3 minutes. We used a mixture of oxygen and nitrogen or CF_4 as the etching gas. Fabricated patterns are observed by scanning electron microscope (SEM, Hitachi, S-4800) after the second RIE.

3. Results and Discussion

When we used our conventional etching condition of second RIE, many resin residues were found on grating patterns. The condition was with the bias power of 100 W, with the reactor pressure of 0.8 Pa, and with the oxygen flow rate of 117 sccm. Chemical composition of the generated reaction products have been analyzed using energy dispersive X-ray spectroscopy (EDX, Horiba, EX-320). EDX spectra after a conventional etching condition are shown in figure 2(b) and 2(c). Oxygen composition in residues area is larger than that in no resin residues area. Thus, we guesse resin residues are oxide-silicon. Figure 3 is a SEM image of grating patterns with a 200 nm pitch and shows there are undesirable residues in the grooves of the gratings. These residues appear when oxygen is exclusively supplied as the etching gas for second RIE. Thus, it is considered that the residues originate in reaction of Si-containing resin and oxygen. We guess that the residues are mainly composed of SiO_2, because the residues are not removed but increased by extenting the etching time of second RIE.

Journal of Physics: Conference Series **191** (2009) 012010 doi:10.1088/1742-6596/191/1/012010

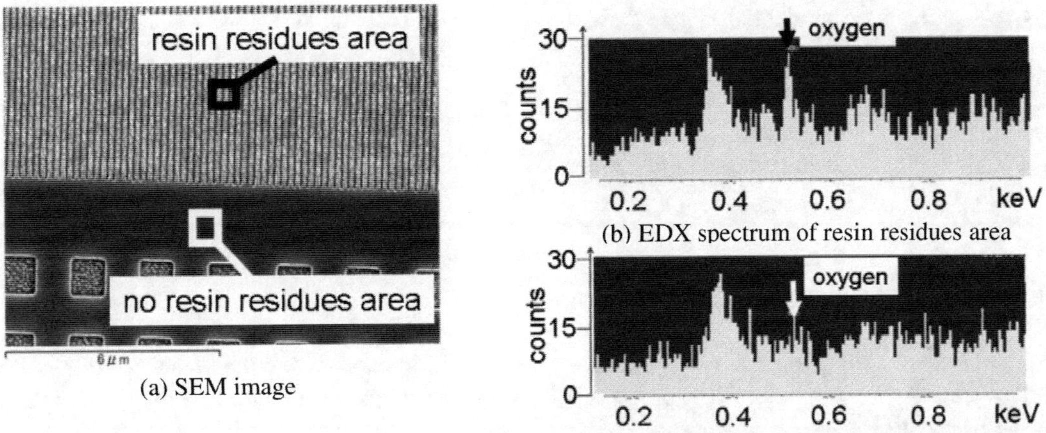

(a) SEM image

(b) EDX spectrum of resin residues area

(c) EDX spectrum of no residues area

Figure 2. EDX analysis after conventional etching condition from the point of (b) resin residues and (c) no residues area which are indicated in (a) SEM

(a) Top view (b) cross sectional view

Figure 3. SEM images of grating patterns with resin residues. Resin residues are shown by a arrow symbols in the photographs.

Figure 4 is a SEM image of cross sectional view of the grating pattern after the second RIE using both oxygen and nitrogen gas. When oxygen flow rate is 4 sccm and nitrogen flow rate is 6 sccm, there are few residues in the grooves. So, it is considered that the reaction between the Si-containing resin and oxygen is suppressed because of less oxygen supplied.

Figure 5 is a SEM image of cross sectional view of grating pattern after the second RIE with the bias power raised to 250 W. It is found that the residues in the grooves are completely disappeared, because residues are mechanically removed by sputtering effect because of the bias power increased.

Figure 6 is a SEM image of cross sectional view of grating pattern after the second RIE with a mixture of oxygen and CF_4. When oxygen flow rate is 50 sccm and CF_4 flow rate is 4 sccm, there are also no residues in the grooves, because the chemical reaction between the resin residues and radicals are increased. However, a large undercut is found in the lower mask layer. In these etching conditions, etching selectivity ratio decreased to more than 20.

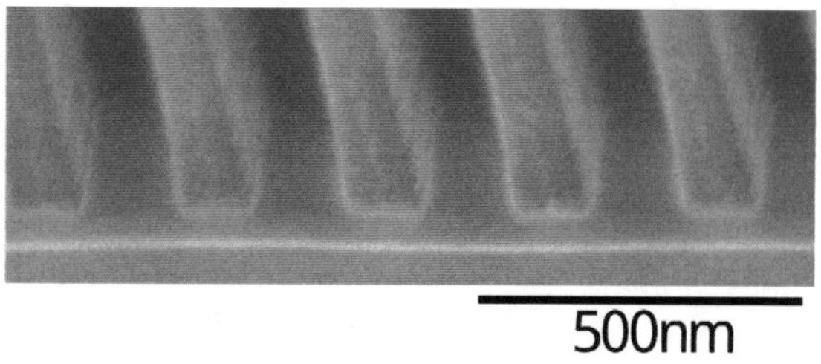

Figure 4. Cross sectional SEM image of grating pattern after the second RIE with $O_2 + N_2$.

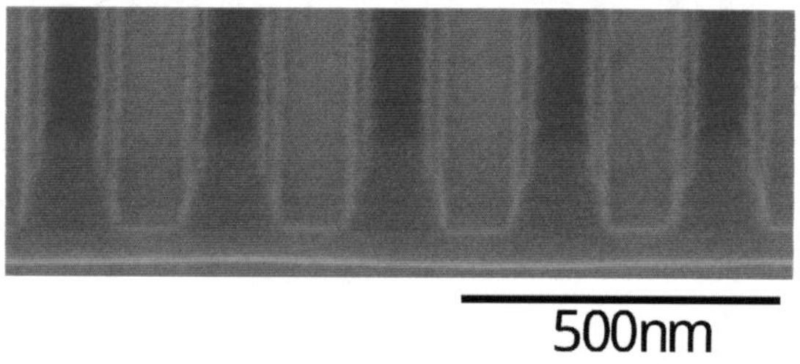

Figure 5. Cross sectional SEM image of grating pattern after the second RIE with $O_2 + N_2$ at 250 W bias power.

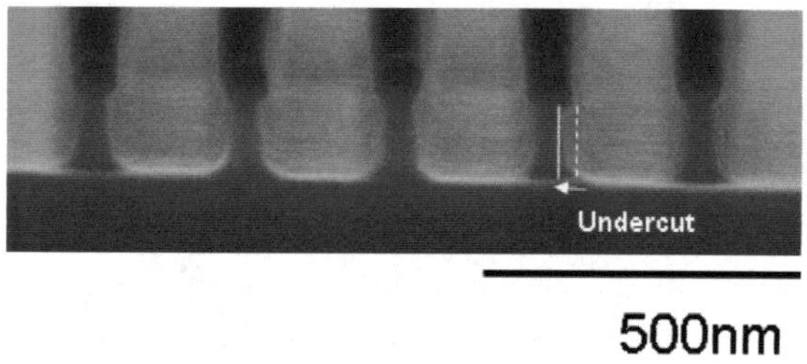

Figure 6. Cross sectional SEM image of grating pattern after the second RIE with $O_2 + CF_4$

4. Conclusion

Several RIE conditions have been examined for transferring nanoimprinted grating patterns to DFB LDs. We have found that the residues are eliminated by using RIE conditions with the mixture of nitrogen and oxygen or CF_4 and oxygen. This is because of the three reasons:
(1) generation of resin residues is suppressed by lowering oxygen ratio for the second RIE, (2) generated residues are removed by sputtering effect due to increasing the bias power of second RIE, and (3) generated residues are removed with enhanced chemical reaction by using a mixture of oxygen and CF_4 gas.
We consider that the large undercut found in (3) condition is undesirable problem, so we conclude that (1) and (2) are the most suitable conditions for fabricating fine grating patterns with the pitch of 200-240 nm.

References

[1] Yanagisawa M, Tsuji Y, Yoshinaga H, and Hiratsuka K 2008 *21st International Microprocesses and Nanotechnology Conference* **29D-9-125** pp 400-401
[2] Miller M, Doyle G, Stacey N, Xu F, Sreenivasan S V, Watts M, and LaBrake D L 2005 *Proc. SPIE* **5751** 994.

Journal of Physics: Conference Series **191** (2009) 012011 doi:10.1088/1742-6596/191/1/012011

Molecular Dynamics Simulation of Nanoindentaiton on Ion-induced Damage of Silicon Surface

S Satake [1], S Yamashina [1],K Ando [1], M Shibahara [2], J Taniguchi [1] and S Momota [3]

[1] Department of Applied Electronics, Tokyo University of Science,2641 Yamazaki, Noda, 278-8510 Chiba, Japan
[2] Department of Mech Eng, Osaka University, 2-1 Yamada-oka,Suita,Osaka 565-0871, Japan
[3] Department of Intelligent Mechanical Systems Engineering, Kochi University of Technology, 185 Miyanokuchi,Tosayamada, Kami, Kochi 782-8502, Japan

satake@te.noda.tus.ac.jp

Abstract. Nanoindentation on ion-induced damage of silicon surface was performed using molecular dynamics. The simulation revealed interaction between a ion-induced damage substrate and a indentation. Two types of silicon substrates are used. One is silicon crystal, and another is amorphous silicon which damaged by argon ion bombardment. The initial velocity of the ion bombardment was 3.807×10^5 m/sec at 30 keV. The computation volume (8.00 nm ×8.00 nm × 17.24 nm) consisted of 57600 Si atoms with Si (100) surface. Each silicon substrate was pressed by the indentation and the force acting on the indentation head were investigated. The force of the ion-induced substrate is smaller than that of the crystalline substrate. Consequently, it turned out that the internal structure of silicon with ion-induced damage can be examined by the force distribution of the indentation.

1. Introduction

Ion beam lithography (IBL) has being rapidly developed as one of nanotechnologies for a semiconductor. Following the development, a superior analytical method is essential now. And the probe of scanning probe microscopes (SPM) such as scanning tunneling microscope (STM) or atomic force microscope (AFM) was introduced and has contributed to development of high accurate analysis for years. Especially nanoindentation test is widely used in field of nanotechnology.

The method introduced by Pharr and Oliver [1] is valid to study material properties in detail. A nanoindentation test needs high accurate control and measurement but is easily put into practice without particular circumstances. Even if measure is quite small, it reveals mechanical properties such as surface, hardness, modulus from just only load-unloading data. Recently, the enhancement of an irradiation effect on the nanohardness on a Si crystal was observed [2], and this enhancement suggests increased defect production in Si crystal by highly-charged ion beams. However, it is difficult to elucidate the mechanism by the experiment because the phenomena caused at the atomic scale order. On the other hand, a molecular dynamics (MD) simulation can solve such as the problem because it can track each ion and atom dynamically. MD simulations about nanoindentation were made by some studies, for instance, a dislocation emission in single crystalline aluminum [3] and phase transformations in silicon monocrystals [4] were presented. In this paper, a crystalline silicon substrate and an ion-induced damaged one were pressed by the indenter. And force acting on the indenter head has been recorded as mathematical values through simulations. The purpose is to investigate each property of the substrates and examine an effect of ion bombardment.

2. The numerical method and the model systems

We performed MD simulations to investigate indentation load-unloading on a Si target. We have previously found an optimal combination of potential functions for the target material and the indentation. The optimal potential combination is found for Silicon, Argon and Carbon : Tersoff potential [5] for the target material and the indenter, Ziegler, Biersack and Littmark (ZBL) potential [6] and Lennard and Jones potential. We used our previous MD code [7]. In the present MD calculations, the Verlet algorithm was used for time integration. Table 1 shows combinations for each potential.

Table 1. Combinations of atoms for potential functions.

Potential function	Pattern
Tersoff	Si-Si,C-C
ZBL	Si-Ar,Ar-Ar
Lennard-Jones	Si-C,Ar-C

2.1. Potential function

For the simulation with a Si target, the Tersoff potential function [5] was employed:

$$\phi = \frac{1}{2}\sum_{i}\sum_{j\neq i} f_c(r_{ij})\left\{f_R(r_{ij}) + b_{ij} f_A(r_{ij})\right\} \tag{1}$$

$$f_R(r) = A\exp(-\lambda r)$$
$$f_A(r) = -B\exp(-\mu r)$$
$$f_c(r) = \begin{cases} 1, & r < R \\ \dfrac{1}{2} + \dfrac{1}{2}\cos\dfrac{\pi(r-R)}{S-R}, & R < r < S \\ 0 & r > S \end{cases}$$

$$b_{ij} = \left(1 + \beta^n \zeta_{ij}^n\right)^{-1/2n}, \quad \zeta_{ij} = \sum_{k\neq i,j} f_c(r_{ik})\, g(\theta_{ijk}), \quad g(\theta) = 1 + \frac{c^2}{d^2} - \frac{c^2}{d^2 + (h-\cos\theta)^2}$$

Here i, j, and k label the atoms of the system, r_{ij} is the length of the ij bond, and θ_{ijk} is the bond angle between bounds ij and jk. The Tersoff potential is also used for Carbon as the indenter. The parameters in Eq. (1) are summarized in Table 2.

Table 2. Tersoff potential parameters for C, Si.[8]

Prameter	C	Si
A[*eV*]	1.3936×10^{3}	1.830×10^{3}
B[*eV*]	3.467×10^{2}	4.7118×10^{2}
$\lambda[Å^{-1}]$	3.4879	2.4799
$\mu[Å^{-1}]$	2.2119	1.7322
β	1.5724×10^{-7}	1.1000×10^{-6}
η	7.2751×10^{-1}	7.8734×10^{-1}
c	3.8049×10^{4}	1.0039×10^{5}
d	4.384	16.217
h	-0.57058	-0.59825
R(1)[*Å*]	1.8	2.7
R(2)[*Å*]	2.1	3.0

2.2. Ziegler, Biersack and Littmark (ZBL) potential.
The Ar ion potential is represented by the Ziegler, Biersack and Littmark (ZBL) potential [6]:

$$\phi(r) = \frac{Z_1 Z_2 e^2}{4\pi\varepsilon r_{ij}}\Phi(r)$$
$$\Phi(r) = \sum_{k=1}^{4} c_k \exp\left(-d_k \frac{r_{ij}}{a_u}\right), a_u = \frac{0.8854a_0}{Z_i^{0.23} + Z_j^{0.23}}, a_0 = 0.529\text{Å} \qquad (2)$$

Here Z_1 and Z_2 are atomic numbers. The detailed parameters are shown in Table 3.

Table3. ZBL potential parameters.

c_i	d_i
0.028171	0.20162
0.28022	0.4029
0.50986	0.94299
0.18175	3.1998

2.3. Lennard and Jones (L-J) potential
Force on Si-C, Ar-C are described in Van der Waals' force by

$$\phi(r) = 4\varepsilon\left\{\left(\frac{\sigma}{r_{ij}}\right)^{12} - \left(\frac{\sigma}{r_{ij}}\right)^{6}\right\} \qquad (3)$$

where r_{ij} is the separation distance between two particles i and j, ε is the potential depth for a parameter of energy, σ is the diameter of an atom. The parameters are shown in Table 4.

Table 4. Lennard-Jones potential parameters.

Combination	ε[eV]	σ[$Å$]
Si-C	2.461	3.555
Ar-C	4.998	3.38

3. Computational parameters

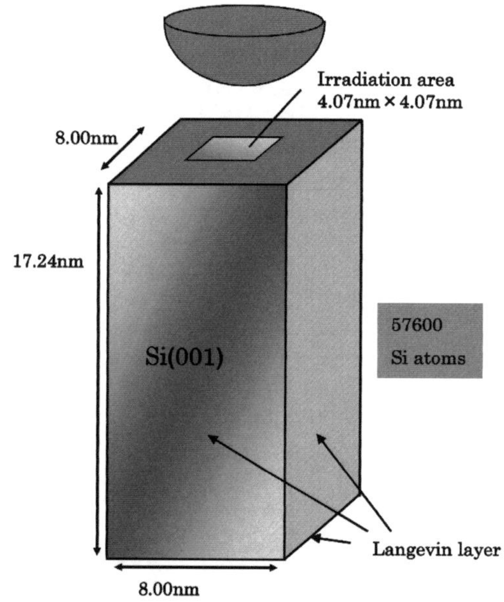

Figure 1. Simulation model.

Computational domain is shown in figure 1. The targets domain was 8.00 nm×8 .00 nm×17.24 nm, consisting of 56,700 Si atoms. Except for the targets domain, the free boundary conditions were applied in the *x*-, *y*- and z-directions. In the targets domain, bottom and all of lateral consist of a fixed heating bath: Langevin layer. The bottom surface is fixed to z-direction. This layer keeps a substrate temperature of 300K. In this simulation two substrates were used. One was single crystalline, the other was amorphous due to Ar ion bombardment with 3.807×10^5 m/sec initial speed at 30keV. These collisions were carried out ten times toward a 4.07 nm×4.07 nm region and incident position is at random and a dose amount was 0.6036×10^{14} ion/cm^2. The computational domain of the substrates was divided into sixteen parallel partitions from the surface to the bottom of the Si target, referred to as regions. A spherical indenter was used and has in total 1269 carbon atoms. The radius was 1.95nm.The indenter moved at a constant velocity of 100m/s. A loading phase continued until the indenter is completely into the substrates, and its move was stopped for 10ps following the previous phase. Then, the indenter was pulled out at the same velocity.

4. Result and discussion

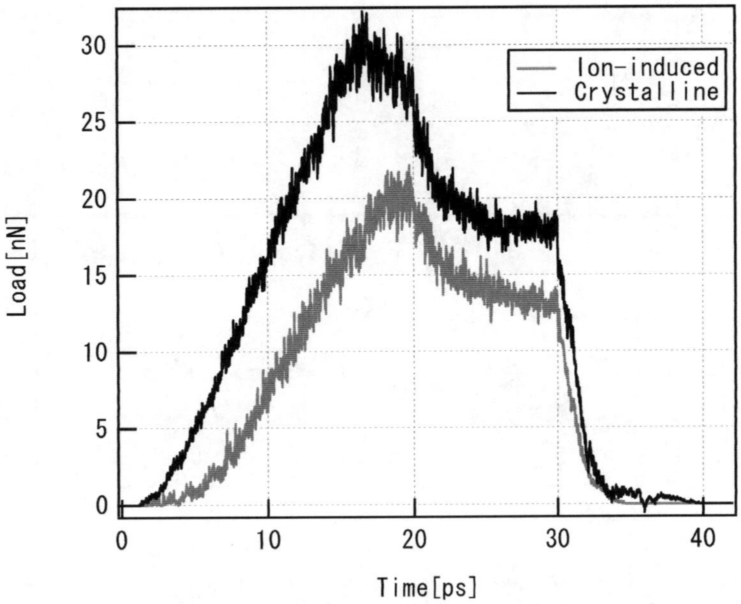

Figure 3. Time dependent of the indent load.

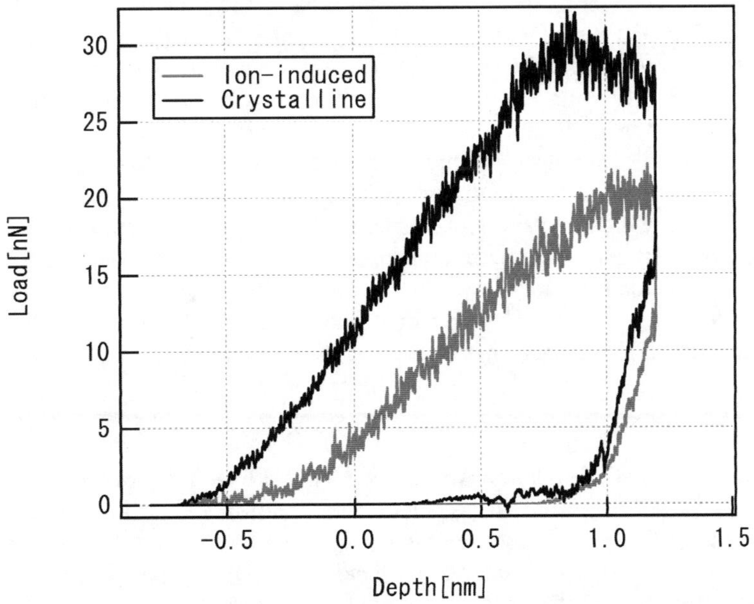

Figure 4. Load-displacement curves.

The result of the simulation about indent load histories is shown in Fig. 3. The force acting on the tip is integrated by the z-components of force acting on the individual atoms during load-unloading process. Figure 3 shows that loads acting on the tip for the crystalline substrate are stronger. In contrast, the histories of the ion-induced substrate are below that of the crystalline one. As a result of the comparison, forces generated on ion-induced state are less than those on crystalline state.

Figure 4 shows load-displacement curves of the ion-induced and crystalline substrate. These curves show that differences between the ion-induced and crystalline substrate are gradually increased as a depth is deeper. Hence it turned out that the ion irradiation affected on the load values.

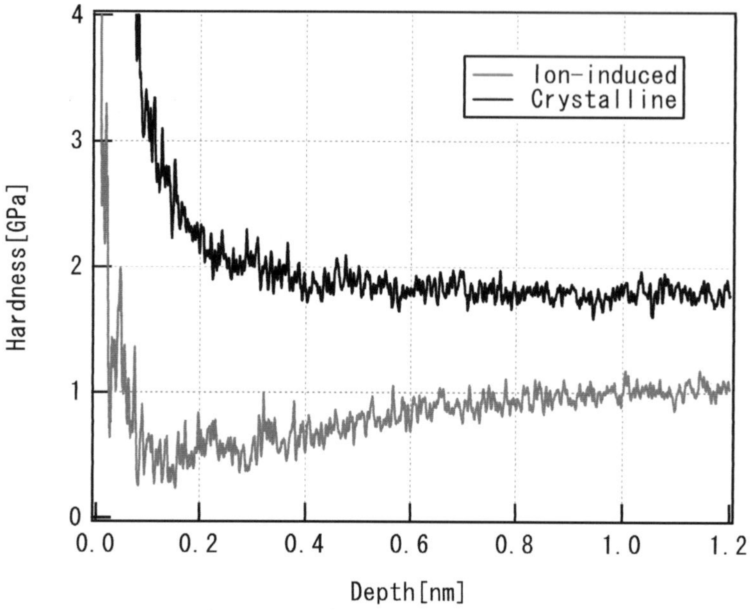

Figure 5. Microhardness of silicon.

Moreover, hardness of the substrates was determined with ***F/A***. ***F*** is value of repulsive forces during the loading. And the ***A*** corresponds to contact area of the indenter. Figure 5 shows calculated results for hardness. The curves of hardness are exponentially decaying as depth is deeper. Both curves have an extreme value around surface. That of the crystalline substrate is broadly flat around 2GPa in 0.2-1.2nm depth. By contrast, that of the amorphous rises slightly in the same range. This phenomenon can be caused by dislocation emission beneath an indenter in a report by Shibutani [9].
In Experimental studies, silicon beneath an indenter transforms from diamond cubic structure to beta-silicon structure at pressure about 8GPa. Amorphous silicon also transforms into beta-silicon structure at 12GPa [10]. In other words, beta-silicon structure is harder than amorphous structure. So the transformation from beta-silicon structure to amorphous structure results in increase of microhardness. In this simulation, the irradiation area is probably on amorphous state in ion collisions. Thus, the hardness of the ion-induced substrate is lower than that of the crystalline. Therefore it has presented that ion-induced substrate is softer than crystalline one.

5. Conclusions

Molecular dynamics simulation was carried out for the nanoindentation on ion-induced damage of silicon surface. The simulation revealed interaction between ion-induced damage and the indentation. Two types of silicon substrates were used. Each silicon substrate was pressed by the indentation and the force acting on the indentation head were investigated. The force of the amorphous substrate is smaller than that of the crystalline substrate. Thus, the ion irradiation affects load and hardness of a substrate. In addition for ion irradiation effect, hardness of large irradiation is softer in that of small irradiation. Consequently, it turned out that the internal structure of silicon with ion-induced damage can be examined by the force distribution of the indentation.

References

[1] Oliver W C and Pharr G M 1992 *J.Mater.Res.* **7** No.6 1564
[2] Pahlovy S A 2008 Ph.D. Thesis, Kochi University of Technology
[3] Shibutani Y and Tsuru T 2004 *The Japan Society of Mechanical Engineers* **70** 947

[4] Cheong W C D and Zhang L C 2000 *Journal of Materials Science Letters* **19** 439
[5] Tersoff J 1989 *Phys. Rev.* B **39** 5566
[6] Ziegler J F, Biersack J P and Littmark W 1985 *The Stopping and Range of Ions in Matter* Pergmon New York
[7] Satake S, Inoue N, Taniguchi J and Shibahara M 2008 *Journal of Physics: Conference Series* **106** 012013
[8] Rafii-Tabar H and Mansoori G A 2003 Interatomic Potencial Models for Nanostructures *Encyclopedia for Nanoscience and Nanotechnology* **X** 11
[9] Tsuru T and Shibutani Y 2004 *The Japan Society of Mechanical Engineers* **73** 738
[10] Cheong W C D and Zhang L C 2000 *Nanotechnology* **11** 173

Journal of Physics: Conference Series **191** (2009) 012012 doi:10.1088/1742-6596/191/1/012012

Surface Deformation of Ion Collision Process via Molecular Dynamics Simulation

S Satake [1], N Inoue [1], S Yamashina [1], M Shibahara [2], J Taniguchi [1]

[1]Department of Applied Electronics, Tokyo University of Science,

2641 Yamazaki, Noda, 278-8510 Chiba, Japan

[2]Department of Mech Eng, Osaka University, 2-1 Yamada-oka,

Suita, Osaka 565-0871, Japan

satake@te.noda.tus.ac.jp

Abstract. Molecular dynamics simulations of Ga ion collision on a Si surface using an optimized potential function were carried out for acceleration voltages of 1, 40, 100 keV for 36, 50, 100 Ga ions. A hillock structure was formed by the Ga ion impact. The height of the structure calculated by the simulations corresponded to experimental values. The sputtering yield was found to be proportional to the acceleration energy and for a high acceleration energy of 100 keV Si atoms deep within the impact surface were sputtered.

1. Introduction

Focused ion beams (FIBs) are a very useful tool in lithography, doping, deposition, and etching. In particular, FIB etching and deposition techniques have been widely used for mask repair and for sample preparation for transmission electron microscopes. Recently, these techniques have been extended to the fabrication of three-dimensional (3D) nanostructures. FIB chemical vapor deposition (CVD) is used to fabricate 3D nanostructures and aerial wiring applied to photonic crystals and nanoscale tweezers. FIB etching, however, has had very limited use in the fabrication and formation of 3D structures, such as predetermined curved shapes, microlens components, and diffractive optical elements, because of the very demanding specifications for etching tasks.

Recently, nanometer-order surface deformation on a Si surface was systematically achieved in experiments with a Ga ion beam collision process after an etch pit was created on the Si surface [1]. The height of the deformation was proportional to the ion beam dose. These experiments demonstrate the requirement for a high-accuracy process technology for nano-masks. The SRIM code [2] has generally been used as a reliable model for ion doping. Although SRIM results are generally highly consistent with experimental data, the detail of sputtered atoms and atomic-order deformation of a surface cannot be tracked dynamically. On the other hand, a molecular dynamics (MD) simulation can track such phenomena, because it can track each ion and atom dynamically. Thus, MD simulation is a powerful and effective tool for analyzing material deformation by ion beam processing.

Previous MD simulations of ion bombardment have studied low acceleration energy [3] and two-dimensional simulations with a two-body potential function [4]. Simulations that assume an actual material through a multi-body potential have been done. However, using time integration with a real radiation time is impossible except for MD simulations of cluster ion irradiation[5] [6][7].

The present study considers nano-order deformation of a surface when the dose value of the irradiation parameter for the calculations is set to be a realistic value for an FIB, while neglecting the time integration problem. Estimations of the deformation height from the simulation results will be important information for the crafting of nano masks by ion beam processing.

Performing this MD simulation is difficult because the interval of the irradiation ions calculated for an actual ion beam facility is of the order of nanoseconds. Therefore, in the simulation the interval of the irradiation ion was compressed and all the ions were assumed to simultaneously collide at the surface. Large-scale MD simulations with the simultaneity condition were performed for acceleration voltages of 1, 40, 100 keV for 36, 50, and 100 Ga ions. The deformation on the surface caused was analyzed after the ions collided on the Si surface.

2. Simulation methods

We performed MD simulations to investigate the Ga ion collision process on a Si target. We have previously found an optimal combination of potential functions for the target material and the impacting ions [8]. The optimal potential combination is found for Silicon and Ga ion: Silicon is Tersoff potential [9], and Ga ion is Ziegler, Biersack, and Littmark (ZBL) potential [10]. The combination indicated the depth being comparable to previous value such as SRIM. We used our previous MD code [8]. In the present MD calculations, the Verlet algorithm was used for time integration.

2.1. Potential function for Si crystal

For the simulation with a Si target, the Tersoff potential function [9] was employed:

$$\phi = \frac{1}{2}\sum_{i}\sum_{j\neq i} f_c(r_{ij})\left\{f_R(r_{ij}) + b_{ij} f_A(r_{ij})\right\} \tag{1}$$

$$f_R(r) = A\exp(-\lambda r) \qquad f_c(r) = \begin{cases} 1, & r < R \\ \frac{1}{2} + \frac{1}{2}\cos\frac{\pi(r-R)}{S-R}, & R < r < S \\ 0 & r > S \end{cases}$$

$$f_A(r) = -B\exp(-\mu r)$$

$$b_{ij} = \left(1 + \beta^n \zeta_{ij}^n\right)^{-1/2n}, \quad \zeta_{ij} = \sum_{k\neq i,j} f_c(r_{ik})\, g(\theta_{ijk}), \quad g(\theta) = 1 + \frac{c^2}{d^2} - \frac{c^2}{d^2 + (h - \cos\theta)^2}$$

Here *i, j,* and k label the atoms of the system, r_{ij} is the length of the *ij* bond, and is the bond angle between bounds *ij* and *jk*. The parameters in Eq. (1) are summarized in Table 1.

2.2. Ga ion for Ziegler, Biersack, and Littmark (ZBL) potential

The Ga ion potential is represented by the Ziegler, Biersack, and Littmark (ZBL) potential [10]:

$$\phi(r) = \frac{Z_1 Z_2 e^2}{4\pi\varepsilon r_{ij}}\Phi(r)$$

$$\Phi(r) = \sum_{k=1}^{4} c_k \exp\left(-d_k \frac{r_{ij}}{a_u}\right), a_u = \frac{0.8854 a_0}{Z_i^{0.23} + Z_j^{0.23}}, a_0 = 0.529\text{Å} \tag{2}$$

Here Z_1 and Z_2 are atomic numbers. The detailed parameters are shown in Table 2.

Journal of Physics: Conference Series **191** (2009) 012012 doi:10.1088/1742-6596/191/1/012012

Table 1. Parameter for Silicon.

	Si
A[eV]	1.8308×10^{3}
B[eV]	4.7118×10^{2}
λ [Å^{-1}]	2.4799
μ[Å^{-1}]	1.7322
n	7.8734×10^{-1}
β	1.1000×10^{-6}
c	1.0039×10^{5}
d	1.6217×10^{1}
h	-5.9825×10^{-1}
R[Å]	2.7
S[Å]	3.0

Table 2. Parameter for ZBL potential.

	c_i	d_i
1	0.028171	0.20162
2	0.28022	0.40290
3	0.50986	0.94299
4	0.18175	3.1998

3. Computational parameters

The computational domain for the large-scale calculation is shown in Fig. 1. The target domain is 25.3 nm × 25.3 nm × 8.55 nm, consisting of 282,752 Si atoms. The MD simulations under the simultaneity condition were performed for acceleration voltages of 1, 40, 100 keV for 36, 50, and 100 GA ions. Periodic boundary conditions were applied in the x- and y-directions. The top boundary was represented by a free boundary condition. The bottom boundary consists of a fixed heating bath layer. Table 3 shows the irradiation areas and ion numbers. Each colored slice in Fig. 2 corresponds to one of sixteen parallel partitions from the surface to the bottom of the Si target, referred to as regions.

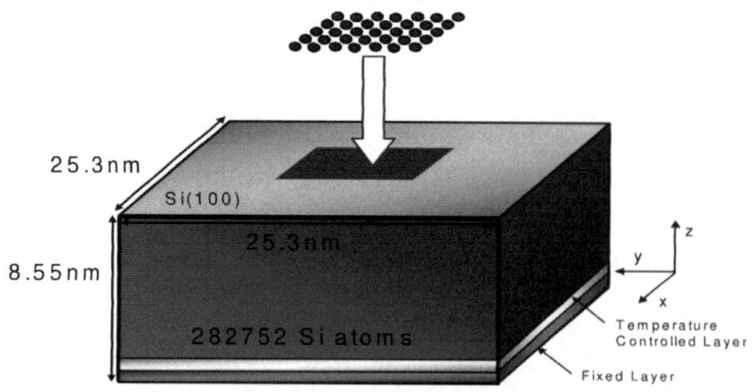

Fig. 1 Computational domain.

Table 3 Parameter for Ga ions.

Number of Ga	Irradiation area	Irradiation density[ions/cm^2]
36	5.430nm × 5.430nm	1.22×10^{14}
50	4.887nm × 4.887nm	2.09×10^{14}
100	4.887nm × 4.887nm	4.18×10^{14}

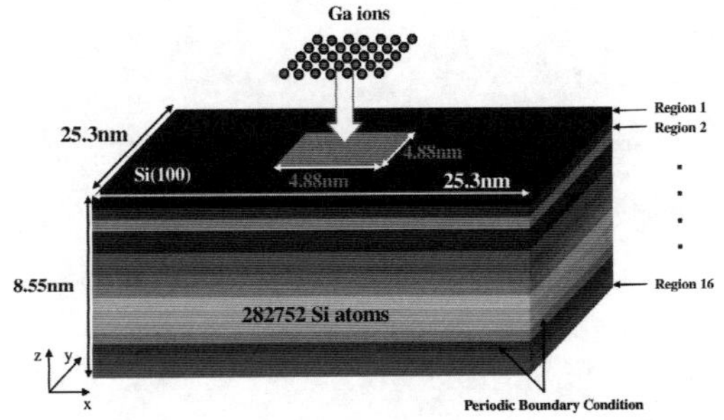

Fig. 2 Computational domain for visualization

4. Results and discussion

Figure 3 shows the height against ion irradiation. The height is determined from the extracted atoms with the kinetic energy calculated for all particles and the binding energy at the Si surface being less than 4.7 eV. Moreover, the height at t = 1.5 ps is expressed as the distance from the coordinate before the collision. Figure 3 (a) shows the height versus acceleration energy for various doses corresponding to the irradiation densities in Table 3. The experimental value by Kawasegi et al. (2005)[1] is also shown in Fig. 3, showing the height irradiated by Ga ions when the acceleration is 30 keV for a dose of 4.2×10^{14} ion/cm^2. In the present simulation, when the acceleration energy is less than 100 keV, the height ranged from 1 nm to 5 nm. In Fig. 3(b), the height is shown versus the dose value. The figure shows the height of the resultant hillock structure, exhibiting an increasing height with increasing dose. Although the time interval of an irradiation ion is compressed, the height of the hillock structure in the simulation corresponds to the same order as the experimental value. However, the height of the simulation is higher than that of the experiment. A time interval of the irradiation ions in the experiment is of the order of nanoseconds. This time interval is sufficiently long to be cooling at the collision surface. On the other hand, the surface temperature of the simulation can not recover owing to the collision of all the ions simultaneously, and the thermal expansion near the surface remained. Therefore, the simulation value has a higher height than the experiment value.

To examine the effect of acceleration energy, the initial ion velocity is changed for the case of 100 ions. The time development at the center surface is investigated by visualizing a 1-nm depth in the z-direction, shown in Fig. 4. In the figure, the computational region is divided by color for the parallel CPUs as in Fig. 2. The each colored layers are destructed by the ion collision, the layers are expanded, and it became the hillocks structure. At t = 0.3 ps for 1 keV, after a low density space is formed at Region 4 or Region 5, the upper layer at the surface appears to sputter. The case for 40 keV is similar to that for 100 keV. Figure 5 shows the probability density function, indicating that the Si atoms are further than 20 nm from the initial target height. The highest layer at the initial target height is the most greatly enhanced for all cases; the deeper atoms are not sputtered. However, at 100 keV, atoms in Regions 13 to 16 are sputtered, whereas this is not observed at lower acceleration energies.

(a)

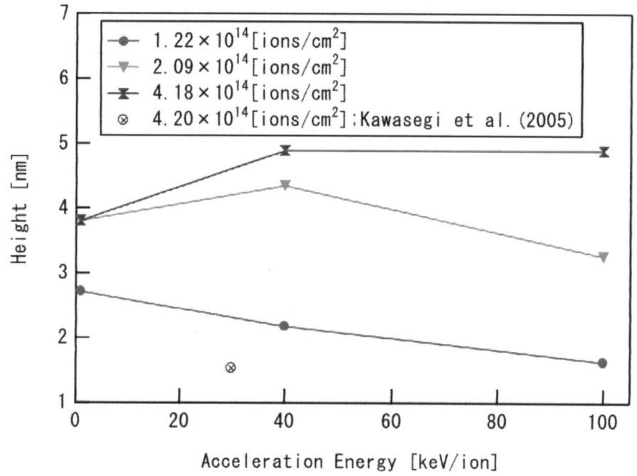

(b)

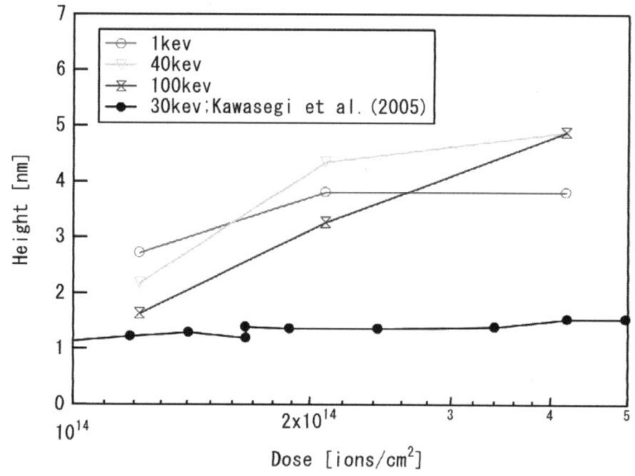

Fig. 3 Height of hillock; (a) Height versus acceleration energy, (b) Height versus dose

(a)

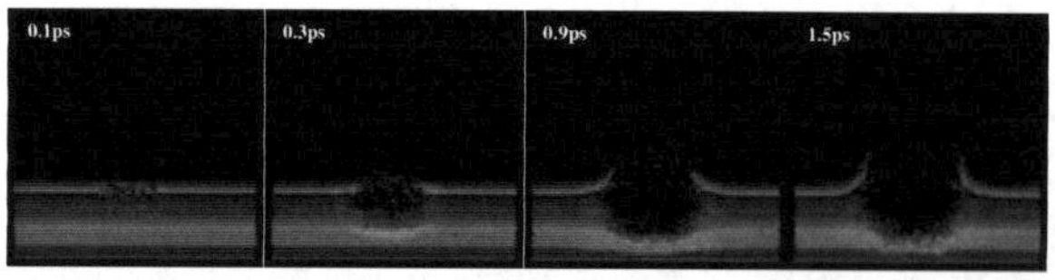

(b)

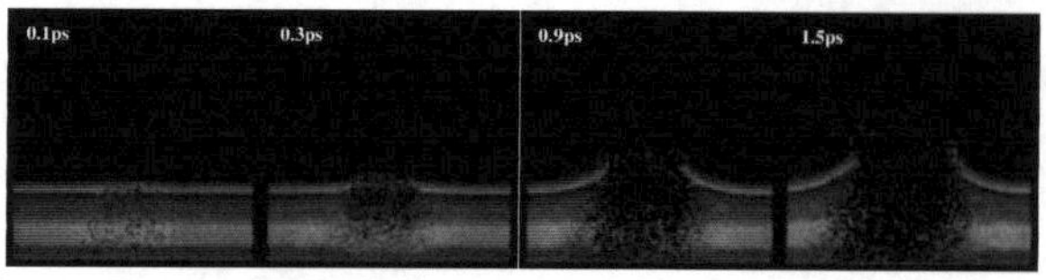

(c)

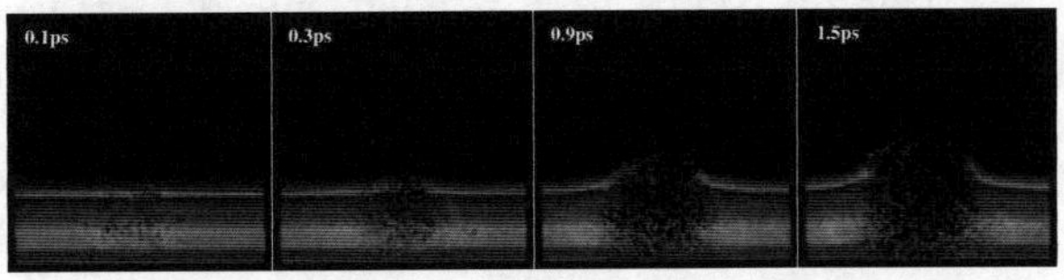

Fig. 4 Snap shot for 0.1, 0.3, 0.9, 1.5ps and the layers are divided by color for the parallel CPUs as in Fig. 2; (a)1keV, (b)40keV, (c)100keV

(a)

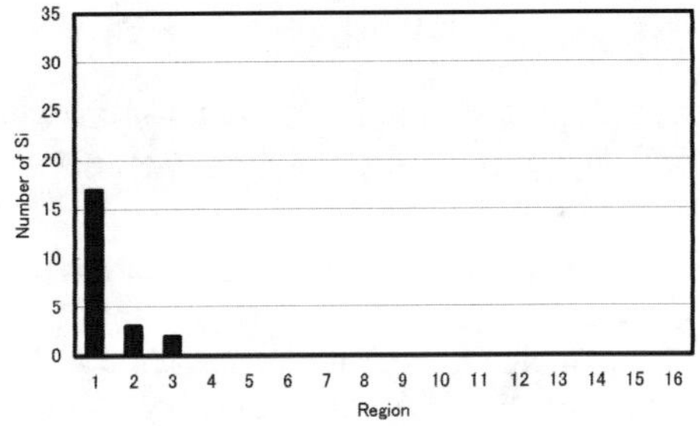

(b)

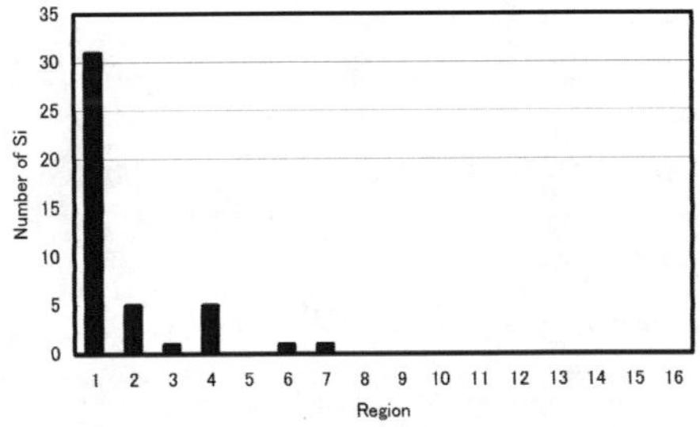

(c)

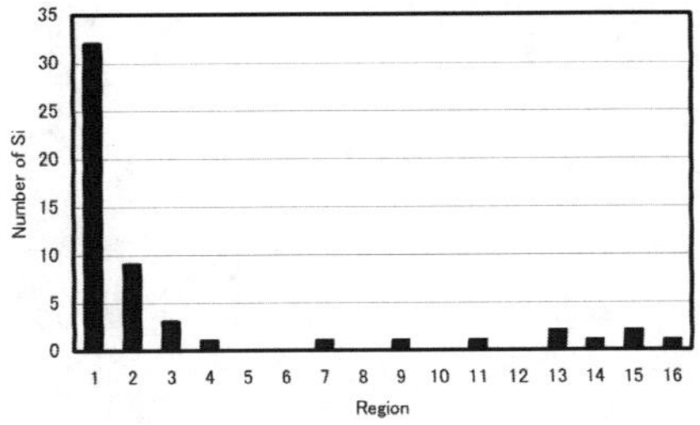

Fig. 5 Number of sputtered atoms and the Regions are divided by color for the parallel CPUs as in Fig. 2; (a) 1keV, (b) 40keV, (c) 100keV

5. Conclusions

Large-scale MD simulations were carried out to understand the influence of dose value and acceleration energy on surface deformation during ion collision. A hillock structure was observed and the height of the structure was analyzed and compared to experimental results.

1. The computation adopted a compressed time interval of ion irradiation because it was impossible to use a real irradiation time. Despite this approximation, the height of the hillock structure is of the same order as the experimental value.
2. The height of the hillock structure is 2 nm when the dose value is 10^{14} ion/cm^2.
3. While the hillock structure is formed, sputtering also occurs. The sputtering yield is proportional to the acceleration energy.
4. For a large acceleration energy, Si atoms deep within the impact surface were also sputtered.

Acknowledgement

This study was supported by "Academic Frontier" Project for Private Universities: matching fund subsidy from MEXT (Ministry of Education, Culture, Sports, Science and Technology), 2006-2010.

References

[1] Kawasegi N, Morita N, Yamada S, Takano N, Oyama T, Ashida K and Taniguchi J Miyamoto I 2005 *Transactions of the Japan Society of Mechanical Engineers C* **71-705** 1754
[2] Ziegler J F *Particle interaction with mater* http://www.srim.org/
[3] Kubota N and Economou D J 1998 *Journal of applied physics*, **83-8** 4055
[4] Müller K H 1987 *Phys. Rev. B* **35** 7906
[5] Postawa Z 2004 *Applied Surface Science* **231-232**, 22
[6] Yamaguchi Y 2002 *Phys. Rev. B* **66** 155408
[7] Satake S Inoue N Kunugi T Shibahara M and Kasahara H 2007 *Nuclear Instruments and Methods in Physics Research B: Beam interact ions with materials and atoms* **257** 639
[8] Satake S Inoue N Taniguchi J and Shibahara M 2008 *Journal of Physics: Conference Series* **106** 012013
[9] Tersoff J 1989 *Phys. Rev. B* **39** 5566
[10] Ziegler J F, Biersack J P and Littmark U 1985 *The Stopping and Range of Ions in Matter* (Pergmon, New York)

Size effect of large deformable nanopillar by focused-ion-beam chemical vapor deposition

H Tanaka[1], M Shinkai[1], Y Shibutani[1,2] and Y Kogo[3]

[1]Department of Mechanical Engineering, Osaka University, 2-1 Yamadaoka, Suita, Osaka 565-0871, Japan.
[2]Center for Atomic and Molecular Technologies, Osaka University, 2-1 Yamadaoka, Suita, Osaka 565-0871, Japan.
[3]Department of Material Science and Technology, Tokyo University of Science, 2641 Yamazaki, Noda, Chiba 278-8510, Japan.

E-mail: sibutani@mech.eng.osaka-u.ac.jp

Abstract. Nanoscopic fabrication technique has been achieved by the direct deposition methods using focused-ion-beam chemical vapor deposition (FIB-CVD). The nanopillar fabricated by FIB-CVD consists of an outer amorphous carbon ring and a inner gallium core. We developed the original double-cantilever (DC) bending test using two pillars rigidly connected by the exposure of a focused electron beam in a scanning electron microscope. The obtained deflection curves suggest that nanopillars have the size dependence to the mechanical response. The pillar with the diameter over 180 nm exhibits a wide region of stiffness weakening after linear response and then becomes extremely hardened at a large deflection. Thus, the pillar intrinsically possesses much more flexibility for bending without any fracturing. The accuracy of a DC testing is also discussed by estimating the bending rigidities of nanopillars, comparing to those obtained by resonance frequency tests.

1. Introduction

Focused ion beam-induced chemical vapor deposition (FIB-CVD) has distinctive advantages, such as much higher deposition rate, in the fabrication of complicated structures. Some delicate microstructures have been actually designed with this micro-beam technology [1,2]. They are composed of the deposited beam members made from amorphous carbon (a-C). Therefore, the overall structural characteristics are, in principle, determined by the mechanical behavior of the member. It has been reported that the a-C pillars have the cross section with a Ga core in nanosized diameter [3]. Thus, the mechanical response of the whole a-C pillar is closely related to the fundamental physical properties as density and Young's modulus, and it has already been investigated by the dynamic resonance vibration and the quasi-static small bending with the piezoactuator [4-6]. We here propose the simple deflection test with a rigid connection of two pillars using a focused electron-beam (EB) in a scanning electron microscope (SEM), and develop the evaluation technique for not the infinitesimal deformation but the large deflection of a-C pillars, which intrinsically equip the remarkable large deformability [7]. In this paper, the bending tests of nanopillars under two conditions are conducted, the parameters of which are the cross-sectional diameter and longitudinal length of a-C pillars, and the size effect of the pillars for large deflection is discussed. In addition, the bending rigidities of pillars are estimated by the results of proposed bending tests, and the accuracy of this test is evaluated by comparing the obtained bending rigidities with that by the resonance vibration tests.

2. Double cantilever test

Nanopillars are grown by a commercially FIB-CVD (SMI9200, Seiko Instruments, Inc.) with the Ga^+ ion beam in the supplied phenanthrene vapor ($C_{14}H_{10}$) precursor gas. In this study, double-cantilever specimens for bending tests are assembled by connecting these two pillars with almost equal cross-sectional diameters (d) and equal longitudinal lengths (l), as shown in figure 1(a). We call it double-cantilever model (DC model). Nanopillars are joined at intersection point of two pillars by irradiation using a focused EB in a SEM (JEOL JSM5310). The residual hydrocarbons from the pump oil (so-called contamination) are absorbed on the pillars' intersection in the process of decomposition by the EB [8,9].

In the DC model, the commercially available AFM cantilever, as shown in figure 1(a), is made of silicon nitride (SiN) coated with Au (BL-RC150VB, Olympus Corporation, Inc.), and it is used in the present bending tests with a spring constant of k=0.006 N/m. The longitudinal length (L), with (b), and thickness (t) of the cantilever are 100(±10), 30(±2), and 0.18(±0.04) μm, respectively. The flexural and tensile rigidities are assumed $E_2I_2 = kL^3/3 = 2.0\times10^{-15}$ Nm^2 and E_2A_2=1.15 N using Young's moduli of SiN and Au (234 GPa and 78 GPa).

We demonstrated large deflection analyses by applying a load to the above DC model, which encompasses the AFM cantilever specified to provide the measurable deflection. This structure seems to be framed rigidly because the connection is exposed to a focused EB for sufficiently long time. The large deflection tests are carried out to displace the AFM cantilever in a vertical direction. The displacement w_1 of the horizontal pillar and displacement w_2 of the AFM cantilever are then measured by in-situ observation using SEM. Figures 1(b) and (c) show SEM images of the bending test. The SEM image, as shown in figure 1(b), is the magnified view of a rigid joint, and the deflection w_1 can be measured. On the other hand, the SEM image, as shown in figure 1(c), is the overall DC testing view, and the deflection w_2 can be measured from this figure. The microstructure was deformed largely beyond a small bending by manipulating the cantilever. Subsequently, it has achieved the tension-dominated shape, in which two members have been linearly oriented (see figures 1(b) and (c)). It is noted that the joint connected by two pillars maintained a right angle even though the joint was subjected to large deflection.

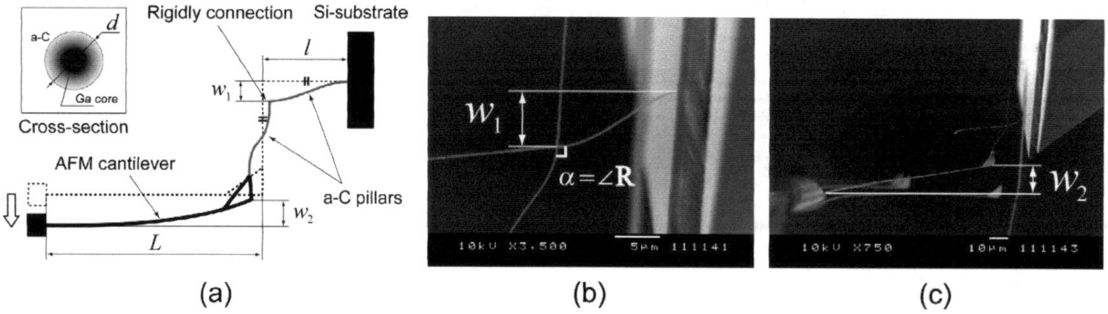

Figure 1. (a) Schematic drawing of a DC bending test, (b), (c) SEM images show the magnified view of a rigid joint and the overall DC testing view.

3. Results and discussion

3.1. Size effect of pillar diameters

Figures 2 show three large deflection curves among three types of pillars (d=160, 180 and 210 nm, and l=15 μm), which are plotted as the relationship between the normalized deflection of w_1/l and the normalized one of w_2/L. The experiments are conducted by two ways of bending to evaluate the elasticity of a-C pillars, one of which is only loading (open symbols), and the other is loading and then unloading (solid symbols). The large deflection curves indicate the weakening of stiffness in the medium-deformation region after linear response, which is remarkably observed for the diameter over 180 nm (see figures 2(b) and (c)). Thus, the pillar has the size dependence that the bending rigidity tends to decrease after linear response as the diameter increases. At finitely deformation, all pillars become hardened due to the geometrical nonlinearity, because the nonlinear elastic of the pillar suggests that tensile rigidity of the pillar contributes more to the large deflection than the flexural rigidity.

In these experiments, the linear and large (except ultimate) deflection behaviors of pillars conform well to the fitted curves of third-order polynomials: Solid lines are fitted by all experimental data. In addition, the dashed lines are the linear elastic solutions obtained by these fitted curves, and each α^* is the gradient of this line. The linear elastic properties of nanopillars are discussed later.

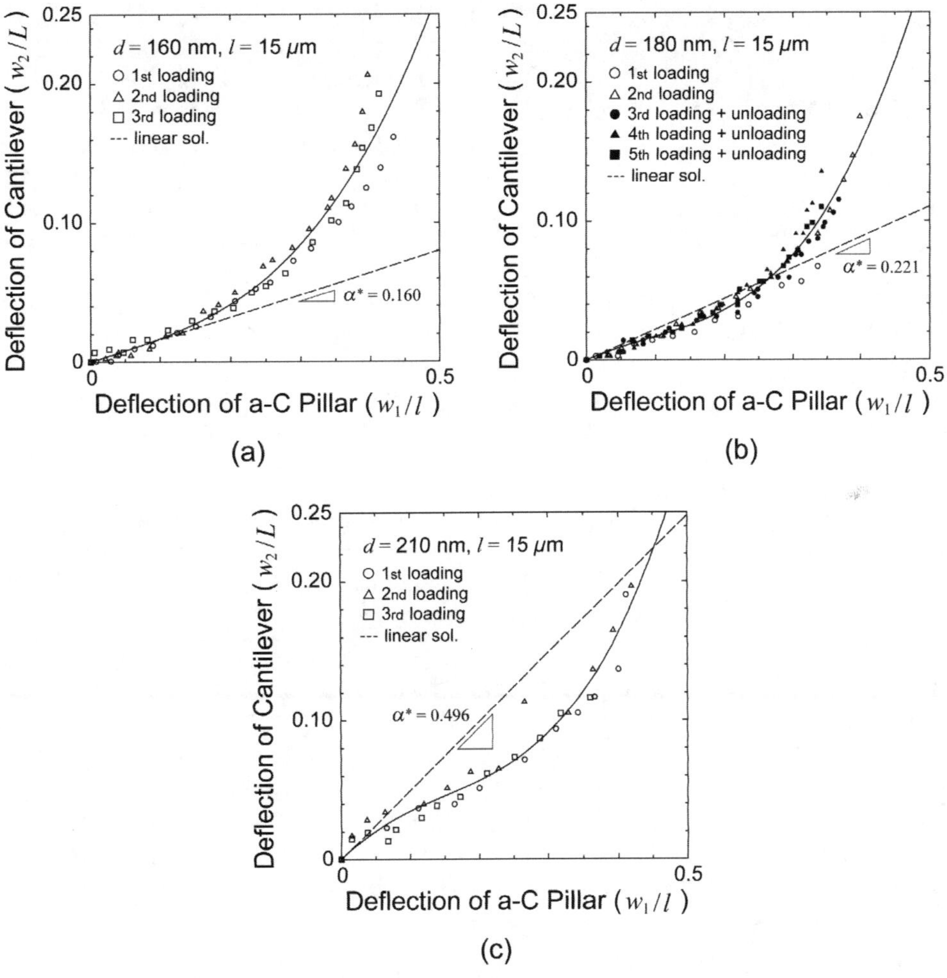

Figure 2. Relationship between normalized deflection of the cantilever and one of the pillar for some diameters (160, 180, and 210 nm). The open symbol represents an experimental value for only loading, and the solid one represents the value of a loading and unloading test. The solid curve here is the cubic polynomial through the origin, fitted by all experimental data. The dashed line is the linear elastic solution obtained by the fitted curve, and α^* is the gradient of this line.

3.2. Size effect of pillar length

We next conduct the DC test by applying the different lengths of pillars with same diameter. Figure 3 shows three large deflection curves among three types of pillars (d=220 nm, and l=15, 20 and 25 μm), and it indicates that nanopillars with short lengths exhibit the same weakening of stiffness as shown in figures 2(b) and (c).

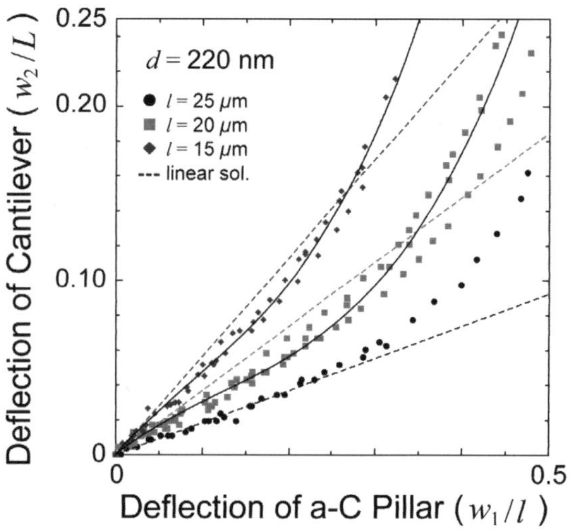

Figure 3. Relationship between normalized deflection of cantilever and one of pillars (d=220 nm and l=15, 20, 25 μm). The solid curve is the cubic polynomial through the origin, fitted by experimental data. The dashed line is the linear elastic solution.

All these results indicate that nanopillars with the diameter within a range of 180~220 nm have loss of stiffness for large deflection, which depends on a ratio of diameter (d) to length (l). This is partly because the internal structure of the a-C pillar changes due to the atomistic deformation manner with covalent interaction based on sp^2-sp^3 bonding. It is sure that the change in a binding mode influence the local density of the pillar, and it express the material nonlinearity in the process of deformation.

3.3. Estimations of the linear elastic properties of nanopillars

When the DC model deforms infinitesimally, the analytical solution derived from the elementary beam theory can afford two Young's moduli E_1 and E_2: E_1 is Young's modulus of the a-C pillar and E_2 is that of the AFM cantilever. The parameter E^*, which is defined by $E^*=E_1I_1/E_2I_2$, can be derived from the elementary beam theory, as follows:

$$E^* = -Z + \sqrt{Z^2 + \frac{1}{2}L^{*3}\alpha^*},\ Z = \frac{1}{16}\left[10L^* - 3L^{*2} + \left(3L^* - 14L^{*2}\right)\alpha^*\right], \quad (1)$$

where $L^*=l/L$, and α^* is equal to w_2l/w_1L and represents the slope in the infinitesimal region of large deflection curve, as shown in figure 2 and figure 3. The flexural rigidity E_2I_2 of the commercial AFM cantilever used in the DC model is the known value. Therefore, the flexural rigidity E_1I_1 of a-C pillars can be obtained from equation (1). Figure 4 compares the obtained flexural rigidities from the present bending tests with those from the previous resonant vibration testing [7]. The solid circle corresponds the bending rigidity derived from the resonance frequency f, as follows:

$$f = \frac{\gamma_n^2 d}{2\pi l^2}\sqrt{\frac{E_1}{16\rho}}, \quad (2)$$

where γ_n is the coefficient of the lowest resonance mode. The details of equation (2) are described in reference [7]. These flexural rigidities for samples with diameters 160 to 220 nm agree well with the predicted value of resonant vibration frequency. Therefore, the proposed DC testing seems to have high accuracy for large deformation region.

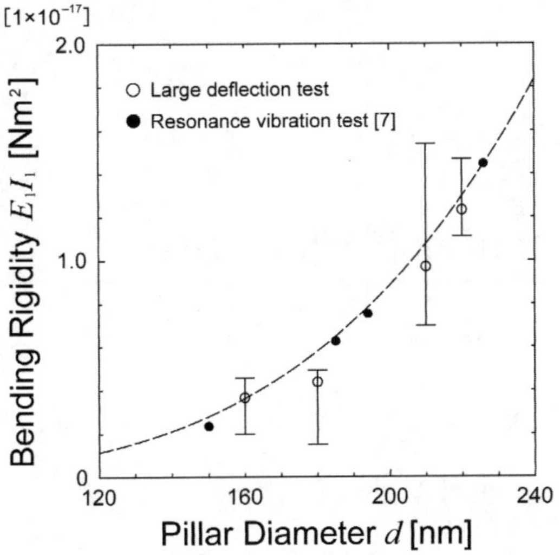

Figure 4. Comparison with the predicted flexural rigidity of resonance frequency: (solid circle) the flexural rigidity E_1I_1 was obtained from the results of the resonance vibration test [7], (open circle) E_1I_1 was calculated the primary coefficient α^* of the fitting equation.

4. Conclusions

We have proposed the experimental technique called double cantilever test to measure large deflection of the pillar manufactured by FIB-CVD. Two identical pillars with the same diameter and the same length are rigidly jointed at the interconnected point by the focused electron beam. Results from the proposed testing verified the size dependence of the deposited pillar to the mechanical response. The pillar exhibits weakening of stiffness in the medium deformation region after linear response and this material nonlinearity depends on a ratio of pillar's diameter to length. Some considerable factors of the stiffness weakening are the properly atomistic deformation manner with covalent interaction based on sp2-sp3 bonding and also the density of the pillar.

Acknowledgments

This author (YS) gratefully acknowledges support from the Grants-in-Aid from Japan Society for the Promotion of Science (Exploratory Research, 20656020).

References

[1] Matsui S, Kaito K, Fujita J, Komura M, Kanda K and Haruyama Y 2000 *J. Vac. Sci. Technol.* B**18** 3168

[2] Matsui S 2007 *Nucl. Instrum. Methods Phys. Res.* B**257** 758

[3] Morita T, Kometani R, Watanabe K., Kanda K, Hoshino T, Kondo K, Kaito T, Ichihashi T, Fujita J, Ishida M, Ochiai Y, Tajima T and Matsui S 2003 *J. Vac. Sci. Technol.* B**21** 2737

[4] Fujita J, Ishida M, Sakamoto T, Ochiai Y, Kaito T and Matsui S 2001 *J. Vac. Sci. Technol.* B**19** 2834

[5] Ishida M, Fujita J, Sakamoto T, and Ochiai Y 2002 *J. Vac. Sci. Technol.* B**20** 2784

[6] Fujita J, Ishida M, Ichihashi T, Ochiai Y, Kaito T and Matsui S 2003 *Nucl. Instrum. Methods Phys. Res.* B**206** 472

[7] Shibutani Y and Yoshioka T 2008 *J. Vac. Sci. Technol.* B**26** 201

[8] Koops H W, Kretz J, Rudolph M., Weber M, Dahm G and Lee L 1994 *Jpn. J. Appl. Phys.* **33** 7099

[9] Nishio M, Sawaya S, Akita S and Nakayama Y 2005 *J. Vac. Sci. Technol.* B**23** 1975

Journal of Physics: Conference Series **191** (2009) 012014 doi:10.1088/1742-6596/191/1/012014

Three-Dimensional Metal Nanoimprint Technique for Electrode and Electric probe

N Unno[1 2], J Taniguchi[1]*, S Ide[1], S Ishikawa[1], Y Ootsuka[3], K Yamabe[3] and T Kanbara[3]

[1] Tokyo University of Science, Department of Applied Electronics
2641 Yamazaki, Noda, Chiba 278-8510, Japan
[2] Research Fellow of the Japan Society for the Promotion of Science
6 Ichibancho, Chiyoda-ku, Tokyo 102-8471, Japan
[3] TIMS and Graduate school of Pure and Applied Sciences, University of Tsukuba
1-1-1 Tennoudai, Tsukuba 305-8571 Ibaraki, Japan
*E-mail: junt@te.noda.tus.ac.jp

Abstract. In the next generation, a strong need exists for nano-scale electrode and electric probe for measuring the electrical characteristics of very small material, such as organic semiconductor. Liftoff technique or etching process with electron beam lithography (EBL) is generally used for fabricating the nano-scale electrode. Although EBL is able to fabricate the sub-10 nm gap pattern, its throughput is not enough to satisfy a demand for mass-producing. Moreover, it is troublesome to fabricate the three-dimensional metal pattern using these processes. In this study, three-dimensional metal nanoimprint technique with metal oxide release layer was examined. As a result, three-dimensional gold pattern was obtained in one shot on the polyethylene terephthalate substrate by using the spin on glass mold coated by Cr_2O_3.

1. Introduction

In the next generation, a strong need exists for nano-scale electrode and electric probe for measuring the electrical characteristics of very small material, such as organic semiconductor. The organic semiconductor device [1], which is manufactured on plastic substrates, has gotten a lot of attention recently because it is flexible, light and inexpensive. In order to operate the organic semiconductor device speedy and integrate densely, the fine metal wiring technique on plastic substrates is required. However, the conventional fine metal patterning technique used for silicon process, such a liftoff technique or etching, is difficult to apply to the plastic substrate due to heat and solvent tolerance. Inkjet printing is a popular and promising process; however, it is difficult to obtain a resolution of less than 1 μm because of the limitation of the ink jet droplets size and the ink bleeding on the substrate. In contrast, nanoprint lithography and nanotransfer printing [2, 3] are appealing processes for the fine metal patterning on plastic substrates. Nevertheless, a resolution of both printing techniques using the polydimethylsiloxane stamp, which is easily deformed, is less than approximately 100 nm for fabricating gap pattern. Moreover, it is troublesome to fabricate the three-dimensional (3D) metal pattern using these processes. Thus, in order to obtain the fine 3D metal pattern easily, we investigate the 3D metal nanoimprint technique using hard mold with metal oxide release layer. The hard mold, which is made by spin on glass (SOG), was fabricated by control of acceleration voltage electron beam lithography (CAV-EBL) [4] and Cr_2O_3 was used for a release layer. As a result, three-dimensional gold pattern was obtained in one shot on the polyethylene terephthalate substrate by using the spin on glass mold coated by Cr_2O_3.

2. Experimental setup

Accuglass-512B (made by Honeywell Co.) which is normally used as interlayer dielectrics was used for positive tone electron beam resist in this study. A buffered hydrofluoric acid (BHF) solution (50 %, HF 25 cm^3/l, and 40 % NH_4F 30 cm^3/l) was used for developer. The calculated concentration of [HF] was 0.43 mol/l and that of [HF^{2-}] was 0.29 mol/l. ERA-8800FE (ELIONIX Co.), which was customized for the delineate task, was used for EBL system with 20 pA of beam current. The fabrication process of three-dimensional metal nanoimprint mold involved three major steps. At first, Accuglass-512B was spin-coated at 3000 rpm on a cleaned Si substrate and cured at 425°C for 1 hour resulting in a 500 nm film. Secondly, the sample was delineated by electron beam (EB) changing acceleration voltage. Then the EB-exposed area on the SOG film was developed out with BHF solution in 60 s. And finally, a Cr layer having a thickness of around 10–20 nm was deposited on a fabricated 3D mold by using a resistively heated vacuum evaporation system (VPC-260F, ULVAC KIKO Inc.). After coating the mold with Cr, we ventilated the vacuum evaporation system and the surface of Cr is oxidized to Cr_2O_3. Next, the transfer process is shown in Figure 1; gold was deposited on the Cr layer. And then, the hot plate was heated up to 80 °C, and the PET substrate was placed on the hard mold for 30 min. After that, the PET substrate was removed, and the gold pattern was transferred onto the PET substrate. The 3D metal pattern on the PET substrate was observed with scanning electron microscope (SEM).

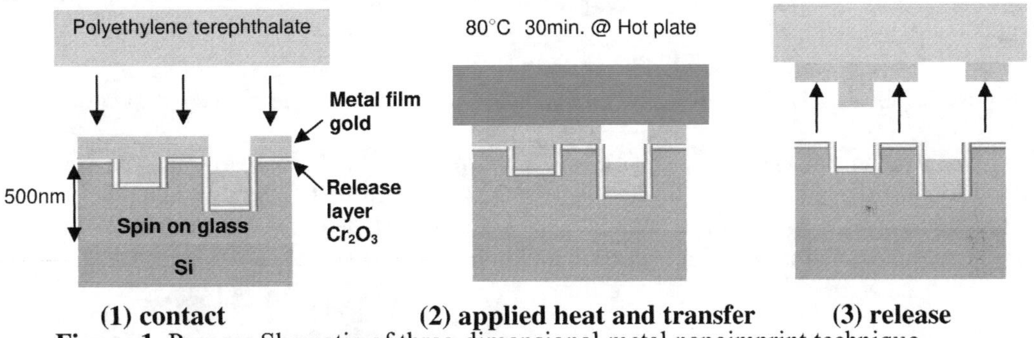

Figure 1. Process Shematic of three-dimensional metal nanoimprint technique.

3. Result and discussion

Figure 2 shows the SEM image of the fabricated 3D metal nanoimprint mold with metal oxide release layer and the 3D metal pattern on the PET substrate. The designed line width and EB dose was 500 nm and 200 $\mu C/cm^2$, respectively. The acceleration voltage was changing from 1 kV to 2 kV. The replicated 3D gold pattern was well corresponded to the mold pattern.

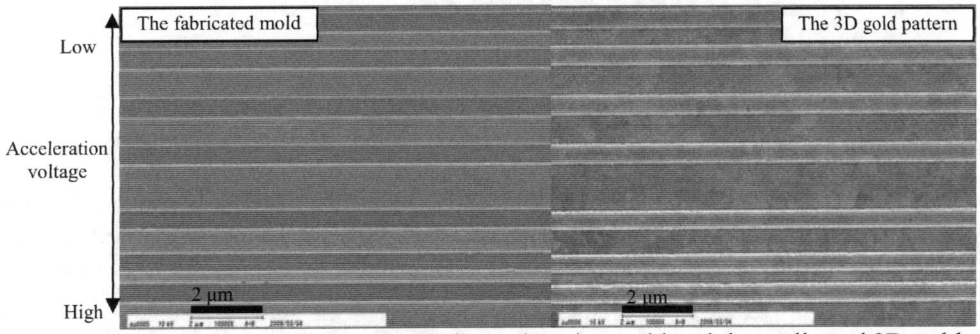

Figure 2. The SEM image of the 3D metal nanoimprint mold and the replicated 3D gold pattern.

Figure 3 shows the atomic force microscope (AFM) image of the mold and replicated pattern same as figure 2. CAV-EBL provides elaborate control of the developed depth. Since the replicated gold patterns were elongated, the replicated gold patterns were higher than the mold patterns because our handmade tool, which was used to release the PET substrate from the mold, lacks the alignment accuracy of parallelism. However, the steps of the replicated pattern heights were approximately equal to the mold pattern.

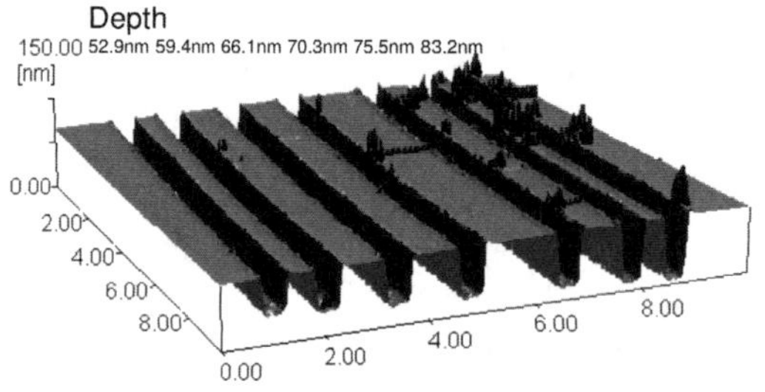

(a)

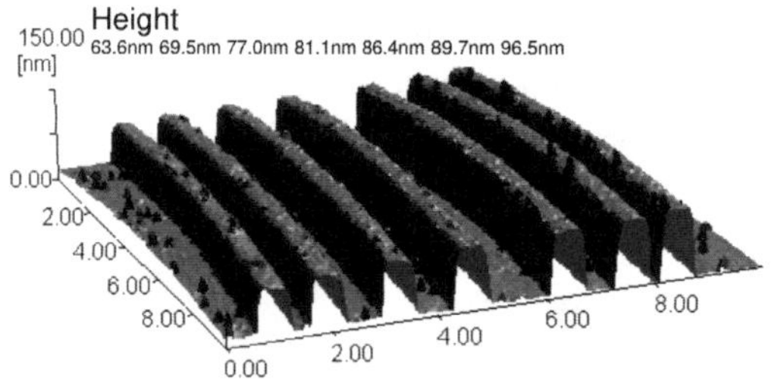

(b)

Figure 3. The AFM image of (a) the fabricated 3D mold and (b) the replicated 3D gold pattern.

Figure 4 shows the concavo-convex shape gold pattern on the PET substrate using the mold which the gold was deposited on the release layer at around 320 nm. This pattern was obtained in one shot since the developed depths of 3 kV, 4 kV were deeper than the deposited thickness of the gold. This technique helps the patterning of the insulating region of nano-scale electrode or probe.

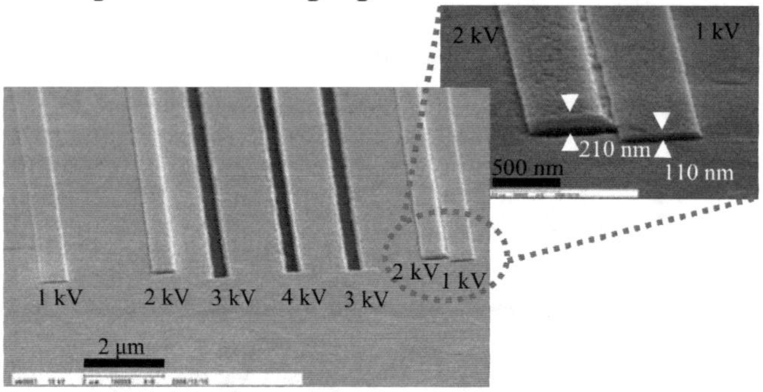

Figure 4. The SEM image of PET substrate after metal nanoimprint by tilting the specimen to 75°.

Figure 5 shows the demonstration of the nano-scale four-terminal electrode on the PET substrate. The thickness of gold was approximately 60 nm. The surface of the gold is very smooth because of the elimination of etching process and it helps easy contact between the sample and the electrode. After 3D metal nanoimprint process, gold remained in the concave area of the mold. However, this process is able to operate repeatedly until gold fills the concave area. And if gold fills the concave area, it is transferred to the PET substrate as a convex pattern, result in cleaning the mold. Therefore, the 3D metal nanoimprint mold with metal oxide release layer is reusable.

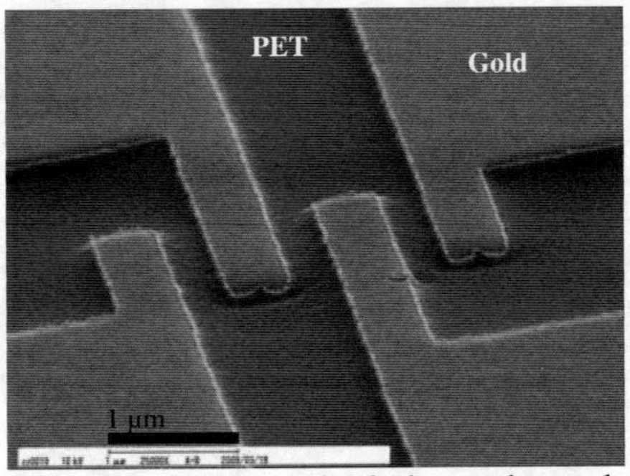

Figure 5. The nano-scale four-terminal electrode on the PET substrate.

4. Conclusion

Three-dimensional metal nanoimprint technique with metal oxide release layer was examined. The three-dimensional gold pattern, which had nano-scale step accuracy, was obtained on the PET substrate. The concavo-convex shape gold pattern was obtained in one shot on the PET substrate and this technique helps the fabrication of the probe shape. Consequently, the nano-scale four-terminal electrode was fabricated. As a result, three-dimensional gold pattern was obtained in one shot on the polyethylene terephthalate substrate by using the spin on glass mold coated by Cr_2O_3. This technique is very useful to fabricate nano-scale electrode and electric probe.

Acknowledgement

This study was supported by "Academic Frontier" Project for Private Universities: matching fund subsidy from MEXT (Ministry of Education, Culture, Sports, Science and Technology) 2006-2010.

References

[1] Yan H, Chen Z, Zheng Y, Newman C, Quinn J R, Dötz F, Kastler M and Facchetti A 2009 *Nature* **457** 679
[2] Kumar A and Whitesides G M 1993 *Appl.Phys.Lett.* **63** 2002
[3] Loo Y L, Willett R L, Baldwin K W and Rogers J A 2002 *Appl. Phys. Lett.* **81** 562
[4] Unno N, Taniguchi J, and Ishii Y 2007 *J. Vac. Sci. Technol.* B **25** 2361

Preparation and electrochemical property of manganese oxide / carbon composites

Y Kameshima*[1], H Wagata, A Nakajima, and K Okada
Department of Metalgy and Ceramic Science, *Tokyo Institute of Technology*
2-12-1 Ookayama, Meguro, 152-8552 Tokyo, Japan

E-mail: ykameshi@ceram.titech.ac.jp

Abstract. A novel electro-chemical capacitor was tried to synthesis by assemble carbon powder with manganese oxide using an ultra-sonication in $KMnO_4$ solution. (ultrasonic sample). A reference sample was prepared by same condition except for radiation of ultra-sonication (settle sample). Both ultrasonic sample and settle sample show the composite of manganese oxide and carbon. Additionally, layered manganese oxide phase was observed in the ultrasonic samples. Surface atomic ratio of Mn/C decreased corresponding with increasing of used solution volume from 25 ml to 250 ml. The composite prepared using 100ml solution was greatly affected by the ultra-sonication. On the other hand, the amount of manganese on the surface has increased with the increasing of the irradiation of energy among the ultrasonic sample. The amount of manganese deposition was almost saturated with 100 mW/cm^3. Cyclic voltammetry curves of the obtained composites were influenced by scanning speed. The ultrasonic sample shows higher electrical capacitance than the settle sample in various conditions.

1. Introduction

Electrochemical capacitor is recently used as a large scale device of starter for diesel-electric engines and hybrid system for electric vehicles.[1)] The electrochemical capacitor shows lower energy density than the secondary battery. Energy capacitor system (ECS), that is a combination of capacitor and electric circuit, is one of the solutions to improve the disadvantage of the electrochemical capacitor.[2)] Shortening the conducting path between the electrode and active material is one effective technique to improve the performance of secondary battery and capacitor. Kawaoka *et al.*[3)] used the activated carbon covered with the manganese oxide as an electrode material of the lithium-ion battery. And, high electrostatic capacitance was obtained by using this electrode under the condition of a high current density and a high-speed electrical charge and discharge. In this work, the manganese oxide/carbon composite that was consisted of manganese oxide covered on the carbon was prepared by introduced the ultrasonic wave irradiation to investigate the electrochemical performance of the resulting composite for electrochemical capacitor.

2. Experimental

2.1. Sample preparation

[1] Present address, Department of Environmental Chemistry and Materials, Okayama University

Ketjen black (KB:EC-600JD(Lion)) were used for the carbon source of the starting material. KB was mixed with distilled water and dispersed using ultra-sonicator of 43 kHz. The $KMnO_4$ solution was added to this carbon dispersed solution under condition of 10 mM $KMnO_4$ solution and carbon/solution ratio was set 0.88 g/l. This solution was irradiated by ultrasonic wave of 28, 45, and 100 kHz for 0.5-24 hours. The solid was dried in 80°C oven after centrifugation and washing by distilled water, and then, manganese oxide/activated carbon composite was obtained. The reference sample was prepared as same process except for the irradiation of ultrasonic wave.

2.2. Characterization

The Crystalline phases in the samples were identified by power X-ray diffractometer (XRD: XRD-6100, Shimazu) with monochromated Cu Kα radiation. Their bulk chemical compositions were determined by energy dispersed X-ray analysis (EDX: EDX Genesis, EDAX) and their surface chemical compositions were determined by X-ray photoelectron spectroscopy (XPS: ESCA-5500MT, PHI). And, the Mn contents in the solution were measured by inductively-coupled plasma optical emission spectrometry (ICP: ICP-OES Prodigy, JEOL).

2.3. Electrochemical measurement

The obtained composite powder mixed with polytetrafluoro-ethylene (PTFE) in the ratio of 10:1, and the mixed powder of 100 mg was formed like the disk of 10mmϕ. Alkali salts were used as an electrolyte for a cyclic voltammetry. A potentio/galvanostat (HABF-5001, Hokuto Denko) was used for the scan of potential. Platinum and a saturated calomel electrode (SCE) were used as a counter electrode and a reference electrode, respectively. A cyclic voltammogram (CV curve) was measured under following conditions; the potential width set from -0.2 to 1.0 V and the scan rate set from 2.5 to 10 mV/s. From the CV curve, the charge-storage capacitance was calculated using the following equation

$$Q = \int \frac{I}{dV/dt} dV$$

where I is the current density [4]. The electrostatic capacitance (C) was estimated by C=Q/ΔV, where ΔV is the width of potential.

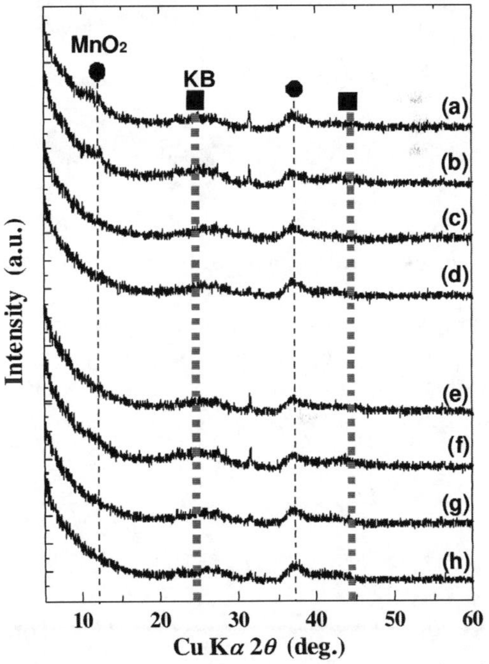

Figure 1. XRD patterns of obtained manganese oxide/activated carbon composites, the Ultrasonic sample (a)250, (b)100, (c) 50, and (d)25 ml and the Settle sample (e)250, (f)100, (g)50 and (h)25 ml.

3. Results and Discussion

3.1. Crystalline phase of composites

Figure 1 shows the XRD patterns of the ultrasonic wave irradiated sample (Ultrasonic sample) and the sample of leaving settle (Settle sample) in various amounts of the solution. Broaden peaks around 2θ=25° and 45° were assigned as halo pattern of carbon from the KB. The peaks around 2θ=12° and 37° were assigned as manganese oxide. The peak around 2θ=12° was only observed in the Ultrasonic sample. This phase was considered to create by ultrasonic irradiation. The layered manganese oxide of Barnesite type has diffraction peak at this position. Therefore, it was shown that the manganese oxide which has high electrochemical performance was synthesized by the ultrasonic irradiation.

3.2. Chemical composition of composite

Figure 2 shows a relationship of manganese/carbon ratio of Ultrasonic sample and Settle sample by XPS and the volume of the using solution. The ultrasonic sample showed a higher Mn/C ratio than the Settle sample on all conditions. Therefore, it is thought that a selectively reduction is occurred on the surface of the carbon by the ultrasonic irradiation. From the ICP analysis, residual Mn content of the Ultrasonic sample was also less than that of the Settle sample. For preparation with 100ml solution, the Mn/C ratio between the Ultrasonic sample and the Settle sample showed the maximum difference.

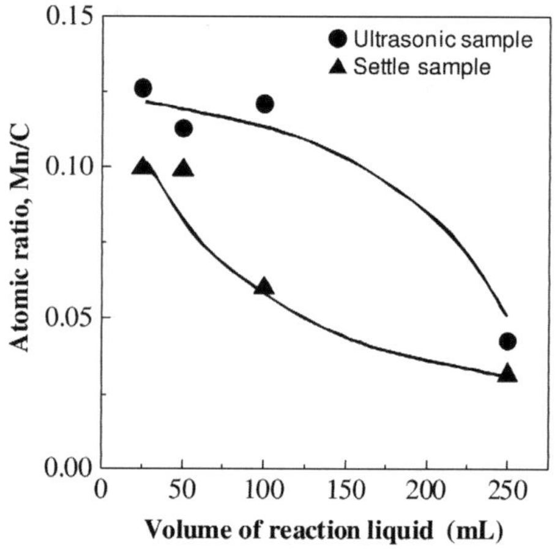

Figure 2. Relation between amount of solution and surface Mn/C ratio of the Ultrasonic sample and the Settle sample.

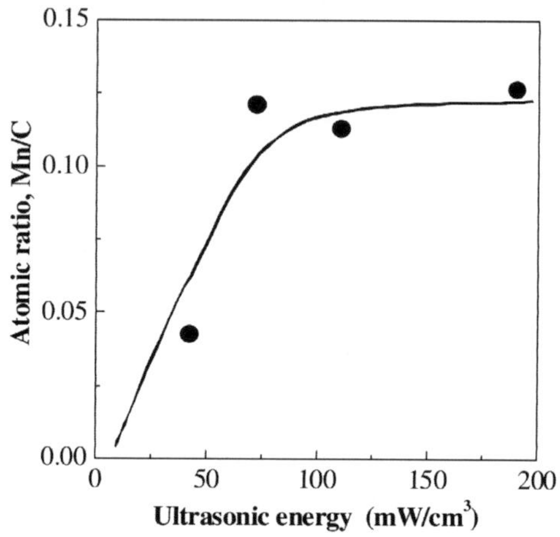

Figure 3. Relation between energy of ultrasonic wave to solution and surface Mn/C ratio.

Figure 3 shows the relationship between the ultrasonic wave energy estimated from the amount of solution and the amount of the surface manganese. Although the amount of the surface manganese increased with increasing of irradiation energy to 100 mW/cm^3, it was saturated over the irradiation energy of 100 mW/cm^3.

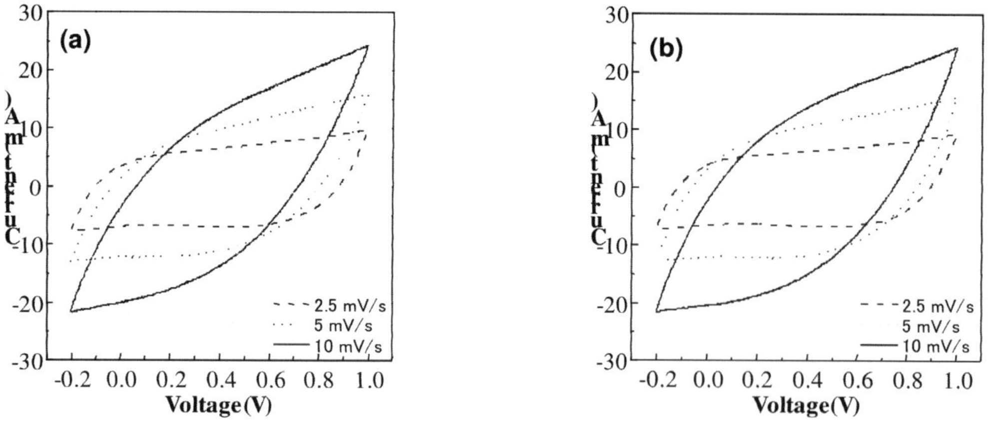

Figure 4. Cyclic voltammogram curves for various scan rate; (a) the Ultrasonic sample and (b) the Settle sample.

3.3. Effect of electrostatic capacitance by an electrolyte

The electrochemical property of the sample prepared with 100 ml solution was evaluated. Figure 4 shows the CV curve in 2M-KCl solution of the Ultrasonic sample and the Settle sample. In each sample, the CV curve has not changed until cycle of 10 times. When the scan rate changed fast, the shape of the CV curve changed narrow hysteresis. This change showed a decrease of the electrostatic capacitance. Figure 5 shows the electrostatic capacitance of each scan rate in the electrolyte of Li_2SO_4, Na_2SO_4, and K_2SO_4. The Ultrasonic sample showed higher electrostatic capacitance than the Settle sample under various scan speed and electrolyte. The maximum capacitance was about 380 F/g of the Ultrasonic sample in Na_2SO_4 or K_2SO_4 for electrolyte. The Ultrasonic sample has a Barnesite type manganese oxide, therefore, Li is easily introduced to interlayer of manganese oxide than Na and K. However, the electrostatic capacitance of the Ultrasonic sample has grown consequentially in order of Li << Na ≈ K. On the contrary, the electrostatic capacitance of the Settle sample has grown consequentially in order of Li ≈ Na << K.

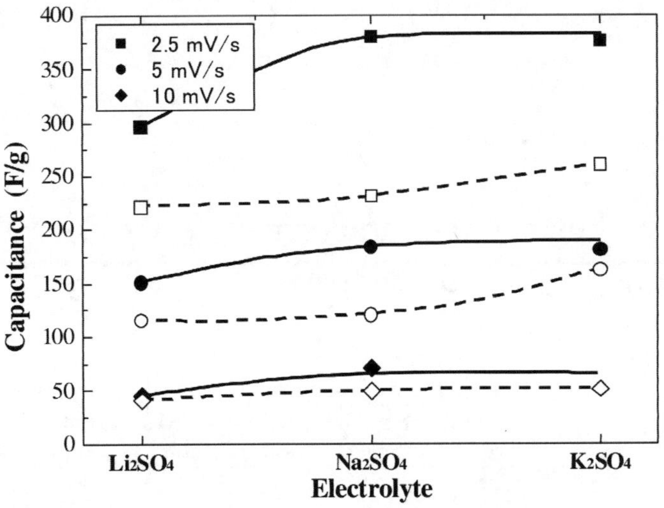

Figure 5. Relationship between electrostatic capacitance and scan speed under various electrolyte; filled: Ultrasonic sample and open: Settle sample.

In generally, the intercalation of K ion into carbon is more effective than that of other alkali ions because the size of K ion is corresponding to the layer distance in the graphite.[5)] Therefore, the contribution of KB to the electrostatic capacitance is larger than that of the manganese oxide in the Settle sample. In the Ultrasonic sample, the effect of K ion was not observed. Therefore, it is considered that the Ultrasonic sample has a structure that the surface is completely covered with the manganese oxide. As a result, manganese oxide/carbon composite, in which carbon is covered by manganese oxide, is obtained by ultrasonic irradiation under reaction.

4. Conclusion

Manganese oxide/carbon composites were successfully prepared by inducing of the supersonic treatment and the following results were appeared:

- The composite, the manganese oxide covered homogeneously on the carbon, was obtained by irradiating the ultrasonic wave of 100 kHz, and the phase of manganese oxide was assigned as Barnesite structure.
- The surface manganese amount of the Ultrasonic sample was higher than that of the Settee sample and especially remarkable difference was observed in the condition preparation using 100 ml solution.
- When the electrostatic capacitance was measured by a cyclic voltammetry, the Ultrasonic sample showed higher electrostatic capacitance than that of the Settle sample.

Acknowledgment
Part of the work was financially supported by the Murata Science Foundation.

References

[1] **Conway B E 1999** *Electrochemical Supercapacitors* **(New York: Kluwer)**
[2] Okamura N 1995 *Denki-gakkai Ronbunshi* **115B** 504
[3] Kawaoka H, Hibino M, Zhou H, Honma I 2004 *J.Power Sources* **125** 85
[4] Pang S-C, Anderson A A, Chapman T W 2000 *J. Electrochem. Soc.* **147** 444
[5] Takeuchi M, Maruyama T, Koike K, Mogami A, Okamura M 1998 *Electrochemistry* **66**[12]

Journal of Physics: Conference Series **191** (2009) 012016 doi:10.1088/1742-6596/191/1/012016

Synthesis and characterization of CuO/ZnO-doped $(Zr_{0.8}Sn_{0.2})TiO_4$ microwave ceramics

A Ghasemi[1,*], A Ashrafizadeh[1], R Shams Alam[2], A Paesano Júnior[3], V Šepelák[4, 5] A Kaewrawang[1], X Liu[1], A Morisako[1]

[1]Spin Device Technology Center, Faculty of Engineering, Shinshu University, Nagano, Japan

[2]Electroceramic Research center, Malek Ashtar University of Technology, Shahin Shahr, Iran.

[3]Centro de Ciências Exatas, Departamento de Física, Universidade Estadual de Maringá, Maringá, Paraná, Brazil

[4]Institute of Physical and Theoretical Chemistry, Braunschweig University of Technology, Braunschweig, Germany

[5]On leave from Slovak Academy of Sciences, Košice, Slovakia

E-mail: ali13912001@yahoo.com

Abstract. The $(Zr_{0.8}Sn_{0.2})TiO_4$ material (ZST), has been prepared by solid state reaction and characterized. The samples were sintered in the temperature range of 1050–1300 °C for 2 h. The effects of sintering parameters like sintering temperature and CuO addition (1-3 wt. %) on structural and dielectric properties were investigated. Bulk density increases from 4.4 to 5.2 g/cm^3 with the increase of sintering temperature and adding elements. The effect of CuO addition is to lower the sintering temperature in order to obtain well sintered samples with high value of bulk density. The material exhibits a dielectric constant of 38.2 and high values of the $Q.f$ product, greater than 60000, at microwave frequencies. The dielectric properties make the ZST material very attractive for microwave applications such as dielectric resonators, filters, dielectric antennas, substrates for hybrid microwave integrated circuits, etc.

1. Introduction

The growing diffusion of personal computers and satellite communication systems and the constant need for miniaturization provide a continuous driving force for the discovery and development of new materials to perform the same or improved functions with decreased size and weight [1,2]. Dielectric resonators (DRs) occupy a prominent position in the development of miniaturization of microwave components because of its advantage of compactness and ease of matching to the microwave integrated circuit [3,4]. For a material to be used as a dielectric resonator, it must have high dielectric constant ($\varepsilon_r > 25$), very high quality factor ($Q > 3000$ at 10 GHz) and very low dependence of resonant frequency on temperature ($\tau_f < \pm 10$ ppm° C^{-1}). The development of materials based on the ZrO_2–SnO_2– TiO_2 ternary system was motivated by the achievement of very low loss and high dielectric constant and a controlled temperature coefficient of the permittivity [4,5]. Remarkable temperature stability can be achieved by using the compound $(Zr_{0.8}Sn_{0.2})TiO_4$ [6,7]. The resonators made of ZST exhibit an almost zero temperature coefficient of the resonant frequency. Therefore, the ZST materials have been successfully applied for microwave devices like low phase-noise dielectric resonator oscillators (DRO), duplexers, filters, frequency discriminators, etc. [8,9]. But the high sintering temperature of this composition makes it difficult to reach a sufficiently dense structure. Three methods have been used to deal with this problem; 1- using glass additives, 2- chemical processing such as sol-gel method or co-precipitation and 3-using sintering aids as liquid flux [10,11]. The first method was found to effectively lower the firing temperature. However, it decreased the microwave dielectric properties of ZST ceramic. The wet chemical process often required a complicated procedure that would increase the cost and time in fabricating the samples. The third method, used in present research, is expected to be able to reduce sintering temperature and keeps the dielectric properties in desired range by adding combination of two or more oxides. The preparation of ZST materials at low sintering temperatures without additives is difficult. The La_2O_3 addition promotes the grain growth. ZnO, increase the sinterability, but no mechanisms have been proposed for the improved densification kinetics or for their effect on the dielectric loss [12-18].

Aiming at the improvement of low-loss dielectric material with a small temperature coefficient of resonant frequency, the dielectric properties of $Zr_{0.8}Sn_{0.2}TiO_4$ ceramics at microwave frequency have been studied. Small amounts of ZnO and CuO additions were added to obtain well-sintered ceramic samples. Consequently, they showed excellent characteristics at microwave frequencies. The effects of additions on the microwave dielectric properties and the microstructures of $Zr_{0.8}Sn_{0.2}TiO_4$ ceramics were also investigated in this article.

2. Experiments

Samples of $Zr_{0.8}Sn_{0.2}TiO_4$ were synthesized by conventional solid-state methods from high-purity oxide powders of ZrO_2, SnO_2, and TiO_2. The starting materials were mixed according to the desired stoichiometry of $Zr_{0.8}Sn_{0.2}TiO_4$. Then 1 wt% ZnO and various amount of CuO (1–3 wt %) were added as sintering aids. The powders were ground in distilled water for 8 h in a ball mill with agate balls. All mixtures were dried and calcined at 1100°C for 3 h. The calcined powders were then remilled for 10 h with PVA solution as a binder. Pellets 13 mm in diameter and 6 mm thick were formed by uniaxial pressing. After debinding, these pellets were sintered at temperatures of 1050–1300°C for 2 h. The heating rate and the cooling rate were both kept at 5°C/min.

The phase compositions were identified by X-rays diffraction analysis using copper target. The bulk densities of the sintered pellets were measured by Archimedes method. The transmission coefficients versus frequency were measured by Vector Network Analyser. The dielectric constant has been measured by transmission method as proposed by William Courtney [12], which is a modified version of Hakki and Coleman method [13] using a Vector Network Analyser (HP8720A). Here the sample is kept between two copper plates and the microwave is inductively coupled to the samples through the antenna connected to the analyzer by the coaxial cables. The dielectric constant was calculated from the sample dimensions and the resonant frequency of TE_{011} mode, which can be easily distinguished from high-frequency modes. The quality factor was measured by reflection method. The temperature coefficient of resonant frequency at microwave frequency was measured in the temperature range of 25–80°C.

3. Results and Discussion

3.1. Structural and dielectric properties

It is identified from XRD patterns (which are not shown here) that $Zr_{0.8}Sn_{0.2}TiO_4$ exhibits an orthorhombic-type crystal structure. On the contrary when additives such as La, Fe, or Bi are used, some second phases were formed, due to the size difference between the substitution ions and the ions in $Zr_{0.8}Sn_{0.2}TiO_4$. Although ZnO is well known to form boundary phases such as Zn_2TiO_4, the second phase was not observed at the 1-3 wt% level of CuO addition, due to the fact that detection of a minor phase by XRD is extremely difficult. It must be mentioned that above 1400 ° C, $ZrTiO_4$ displays a structure with a random distribution of Zr^{4+} and Ti^{4+} cations and addition of Sn^{4+} would stabilize this

Journal of Physics: Conference Series **191** (2009) 012016 doi:10.1088/1742-6596/191/1/012016

disordered structure. Disorder-order transformation in ZST can be affected by these changes. It has been proved that change in structure from disorder to order state, is accompanied by reduction in unit cell volume. As these two have been increased with CuO addition, it can be concluded that adding CuO, would not accelerate the disorder-order transformation.

The plots of bulk densities of 1 wt% ZnO-doped $Zr_{0.8}Sn_{0.2}TiO_4$ ceramics with different amounts of CuO addition versus sintering temperature are illustrated in Fig. 1. As shown in Fig. 1(a), the $Zr_{0.8}Sn_{0.2}TiO_4$ ceramics were dense at 1250°C and possessed densities higher than 95%. Initially, the density increased with increasing sintering temperature and seemed to saturate at 1250°C. After reaching the maximum at 1250°C, it decreased slightly owing to the appearance of open pores. It seemed that CuO did attribute to the densification of $Zr_{0.8}Sn_{0.2}TiO_4$ ceramics. The maximum relative density was found to be 96.5% with 3 wt% CuO addition at 1250°C. It is difficult to synthesize ceramics with ultrahigh relative density by the conventional solid-state method unless it is augmented by other methods, such as chemical processing.

Figure 1(b) shows the dielectric constants of 1 wt% ZnO-doped $Zr_{0.8}Sn_{0.2}TiO_4$ ceramics with different amounts of CuO addition as functions of their sintering temperatures. As can be seen, the addition as well as the sintering temperature did not significantly affect the dielectric constant. The ε_r values of the well-sintered $Zr_{0.8}Sn_{0.2}TiO_4$ ceramics ranged from 36.2 to 38.2. As dielectric constant of the material is related to dielectric constant of each phase present in the sample and its volume fraction, and porosity is regarded as a separate phase with dielectric constant of 1, increase in density caused by CuO addition, raised the dielectric constant. Another effect is the increase of the c_0-axis length which is correlated to increase in ionic polarization and can lead to higher dielectric constant.

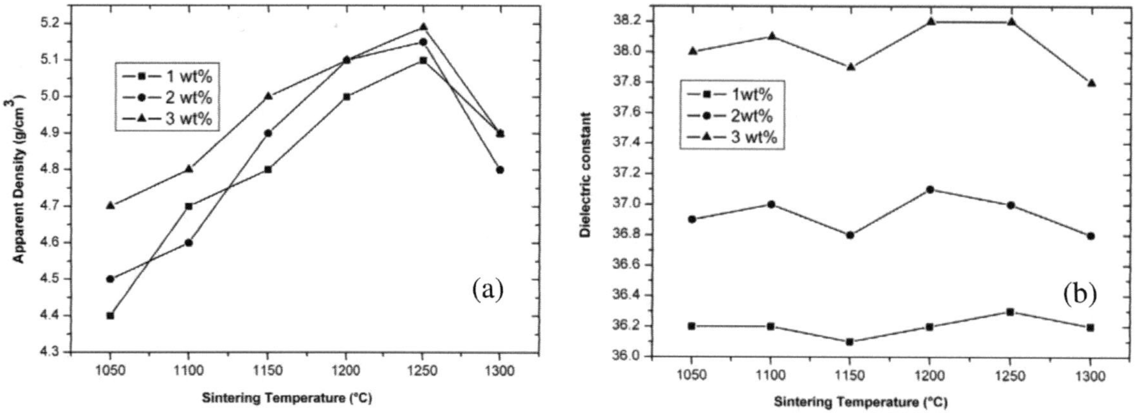

Fig. 1. (a) The relative densities and (b) dielectric constant of ZST ceramics with different amounts of CuO addition versus sintering temperature

3.2. Microwave properties

The variation of transmission coefficients with frequency for samples prepared at 1250 °C are shown in Fig. 2. The sample contains 1wt% CuO exhibits the largest linewidth in compare to the other samples. The maximum transmission coefficient is belonging to resonant frequency of 11.2 GHz. There are another two resonant peaks which might be attributed to the others excitation modes in microwave region. The transmission coefficients of the other excitation modes are more than -20 dB at 8.7GHz and -9 dB at 12.2 GHz respectively. The resonant frequency and transmission coefficient can be easily engineered by adding the additional amount of CuO in the component. It reveals that the sample with 2 wt% CuO provides transmission coefficient value of -33.62 dB at 13.8 GHz. In this material no secondary peaks were found. The largest transmission coefficient along with resonant frequency could be achieved with incorporation of 3 wt% CuO in the microstructure. It is clearly observed that the originated linewidth is negligible. Consequently, it can be concluded that with controlling of CuO additive, the manipulation of high Q.*f* is plausible. The results regarding to the Q.*f* values will be explained in this section.

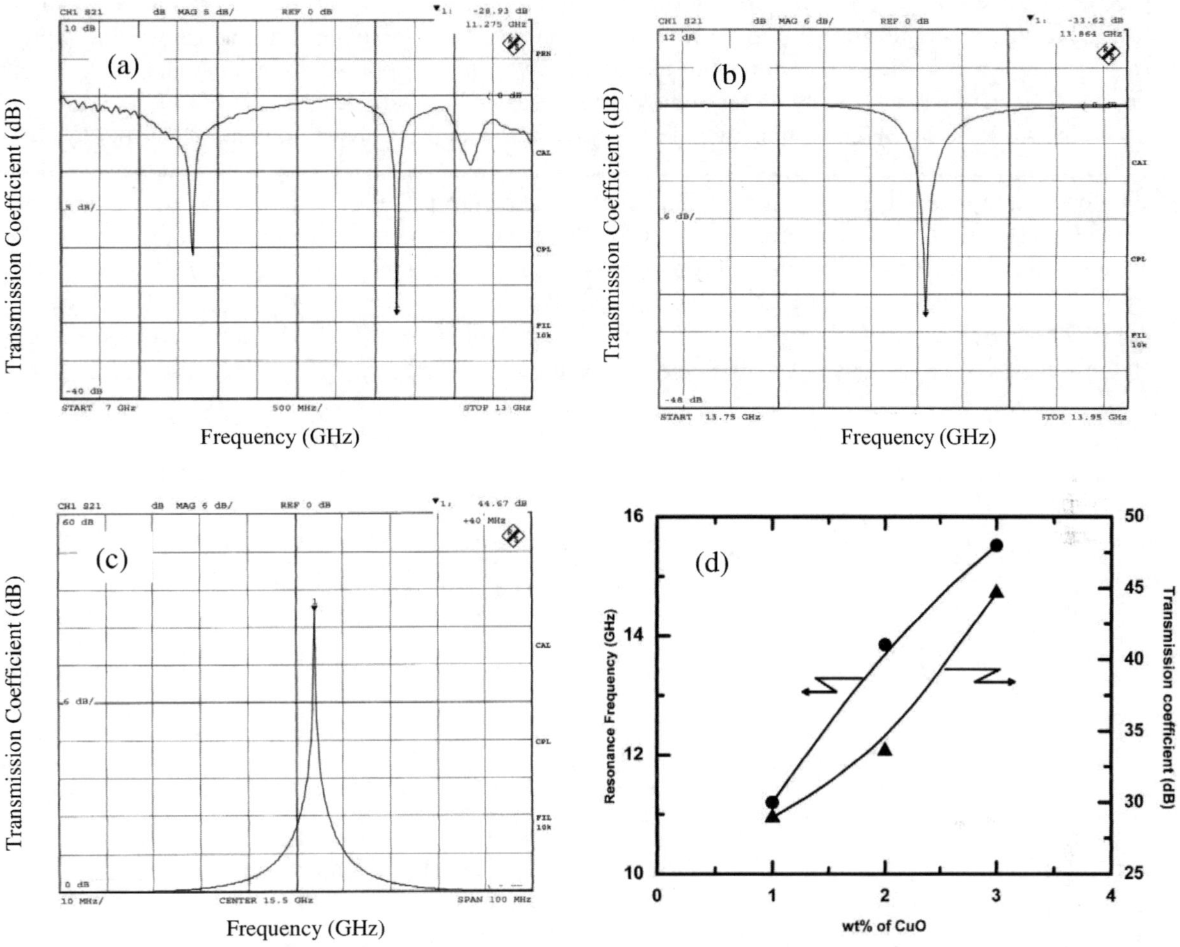

Fig. 2. Transmission coefficient versus frequency of ZST doped with (a) 1 wt% CuO, (b) 2 wt% CuO, (c) 3wt% CuO, and (d) variation of resonant values with amount of CuO

The resonant frequency is known to be related to the length and diameter of cylindrical samples along with dielectric constants. In this research, we treat the dimension as a constant due to the same values of length and diameter in all prepared samples. Consequently the position of resonant frequency is due to the variation of dielectric constants with adding elements. Figure 2(d) clearly reveals the relevant data.

The Q.*f* values of 1 wt% ZnO-doped $Zr_{0.8}Sn_{0.2}TiO_4$ ceramics with various amounts of CuO addition, as functions of their sintering temperatures, are shown in Fig. 3. The relationships between Q.*f* values and sintering temperatures reveal the same trend as those between densities and sintering temperatures. Relative density plays an important role in controlling dielectric loss, as has been shown for other microwave dielectric materials. With 3 wt% CuO addition, the Q.*f* value of $Zr_{0.8}Sn_{0.2}TiO_4$ ceramics increased from 29,000 at 1050°C to 61,000 at 1250°C and then decreased with further increase in the sintering temperature. However, the Q.*f* values decreased with the increase of CuO addition at fixed sintering temperature. It has been reported [6,7] that the microwave dielectric loss is mainly caused not only by the lattice vibrational modes, but also by the pores, the second phases, the impurities, or the lattice defect. Because the second phase was not observed, the degradation in $Q.f$ value might be due to impurities. One can generally state that to obtain a maximum quality factor, it is necessary to produce a perfect single phase material devoid of doping or impurity. For a perfect crystal, the quality factor would be limited only to intrinsic lattice dampening, which depends on crystal structure. However, when densification requires the use of additives, it is important to identify the mechanism involved. Second phases usually increase dielectric losses and oxygen vacancies are important factors on dielectric loss for ZST ceramic. In this research no secondary phase containing Cu or Zn was detected.

The temperature coefficient of resonant frequency is known to be related to the composition and the second phase of the material. Because $Zr_{0.8}Sn_{0.2}TiO_4$ ceramics are temperature stable, and CuO addition did not cause an observable second phase, the τ_f values did not change much and ranged from -3.2 to -1.2 ppm/°C in the experiment (Fig. 3(b)). With 3wt% CuO addition, the 1 wt% ZnO-doped $Zr_{0.8}Sn_{0.2}TiO_4$ ceramics sintered at 1250°C had excellent microwave dielectric properties with an ε_r value of 38.2, a $Q.f$ value of 61,000, and a τ_f value of -1.2 ppm/°C.

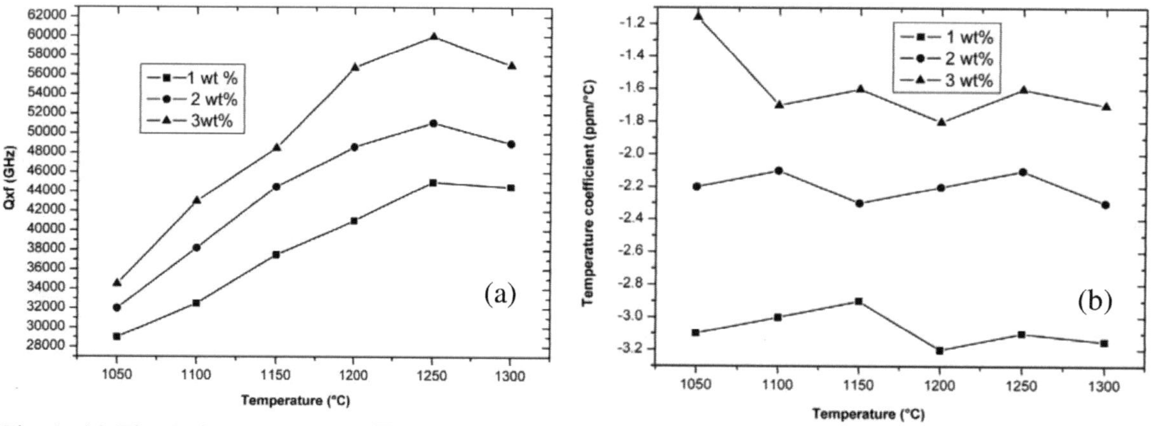

Fig. 3. (a) The $Q.f$ values and (b) τ_f of $Zr_{0.8}Sn_{0.2}TiO_4$ ceramics with various amounts of CuO addition given as a function of sintering temperature.

4. Conclusions

The effect of CuO addition on the microwave dielectric properties and the microstructures of 1 wt% ZnO-doped $Zr_{0.8}Sn_{0.2}TiO_4$ ceramics were investigated. $Zr_{0.8}Sn_{0.2}TiO_4$ ceramics with ZnO and CuO additions can be well sintered to approach 96.5% theoretical density at 1250°C. The $Q.f$ value of $Zr_{0.8}Sn_{0.2}TiO_4$ ceramics can be promoted to be as high as 61,000 at this sintering temperature. The reduction of the dielectric loss of $Zr_{0.8}Sn_{0.2}TiO_4$ ceramics with CuO addition was mainly attributed to enhanced densification. $Zr_{0.8}Sn_{0.2}TiO_4$ ceramics with 1 wt% ZnO and 3 wt% CuO additions sintered at 1250°C exhibited an ε_r of 37.8, a $Q.f$ value of 61,000, and a τ_f value of -1.2 ppm/°C. Compared to other additions, $Zr_{0.8}Sn_{0.2}TiO_4$ ceramics with CuO added exhibited enhanced $Q.f$ values without significantly changing the dielectric constant or temperature coefficient.

Acknowledgment

This work was supported in part by Research (B) from JSPS and Grant-in-Aid for Scientific Research Tokobetstu Kenkyuin Shorei-hi. V.Š. thanks the Deutsche Forschungsgemeinschaft, APVV (Project 0728-07) and VEGA (2/0065/08) for the support of his work.

References

[1] Nishigaki S, Kato H, Yano S and Kamimure R 1987 *Am. Ceram. Soc. Bull.* **66** (9)1405
[2] Wakino K, Minai K and Tamura H 1984 *J. Am. Ceram. Soc.* **67** (4) 278
[3] Christoffersen R, Davies P K and Wei X 1994 *J. Am. Ceram. Soc.* **77** (6) 1441
[4] Schmidt D and Weiland T 1992 *IEEE Trans. Magn.* **28** 1793
[5] Tamura H 1994 *Am. Ceram. Soc. Bull.* **73** 92
[6] Park Y H, Ryu J M, Shin M Y, and Ko K H 2001 *J. Am. Ceram. Soc.* **84** 2542
[7] Zhang S X, Li J B, Zhai H Z and Dai J H 2002 *Ceram. Int.* **28** 407
[8] Yuksei E Y, Nagode T and Wong T T Y 2004 *IEE Proc. Microw. Antennas Prorag.* **151** 221
[9] Leeson D B 1996 *In Proc. IEEE.* **54**, 329
[10] Michiura N, Tatekawa T, Higuchi Y and Tamara H 1995 *J. Am. Ceram. Soc.* **78** 793
[11] Heiao Y C, Wu L and Wei C C 1988 *Mater. Res. Bull.* **23** 1687
[12] Courtney W E 1970 IEEE *Trans. Microwave Theory Tech.* **18** 476
[13] Hakki B W, Coleman P D 1960 *IEEE Trans. Microwave Theory Tech.* **8** 402
[14] Huang C L and Weng M H 2000 *Mater. Res. Bull.* **35** (11) 1881
[15] Huang C L, Weng M H, Wu C C and Wei C C 2001 *Jpn. J. Appl. Phys.* **40** 698
[16] H. Tamura 1994 *Am. Ceram. Soc. Bull.* **73** (10) 92
[17] Silverman B D 1962 *Phys.Rev.* **125** 1921
[18] Kim W S, Hong T H, Kim E S and Yoon K H 1998 *J. Appl. Phys.* **37** 5367

Journal of Physics: Conference Series **191** (2009) 012017 doi:10.1088/1742-6596/191/1/012017

Thin Film Growth Behaviors on Strained fcc(111) Surface by kinetic Monte Carlo

Y Doi [1], D Matsunaka [1,2], and Y Shibutani [1,2]

[1]Department of Mechanical Engineering, Osaka University, Osaka 565-0871, JAPAN

[2]Center for Atomic and Molecular Technologies, Osaka University, Osaka 565-0871,JAPAN

doi@comec.mech.eng.osaka-u.ac.jp

Abstract. We study Ag islands grown on strained Ag(111) surfaces using kinetic Monte Carlo (KMC) simulations. We employed KMC parameters of activation energy and attempt frequency estimated by nudged elastic band (NEB) method and vibration analyses. We investigate influences of surface strain and substrate temperature on film growth. As the biaxial surface strain increases, the island density increases. As temperature increases, the shape of the island changes from dendric to hexagonal and the island density increases.

1. Introduction

Thin film technology has enabled the development of applied microelectronic devices which have various functional properties, such as magnetic, thermal, chemical ones. One of the most widely used techniques which fabricate thin films is Molecular Beam Epitaxy (MBE) [1]. MBE can precisely control the thickness of films, and manufacture nanoelectronic devices. It is important to understand the growth process at the atomic level and its relationship with film characterization.

At heterointerfaces, periodic dislocation network could appear due to the lattice misfit [2-5]. The STM results show that well-ordered islands are formed upon periodic dislocation network [3-5]. The dislocation network gives rise to periodic strain on the film surface. Surface strain often governs the growth behavior. *Ab initio* self-consistent calculations have shown that a Co adatom on the Pt (111) surface prefers to diffuse to the largest tensile strain region [6]. The Strain-assisted self-assembly has been expected to be one of the promising methods to achieve controlled nanostructures from the thermodynamic and kinetic points of view. Thus, it is worthful to study effects of the surface strain on the island nucleation and the growth behavior by atomic simulations.

In this study, we simulate the homo-epitaxial growth of Ag film on strained Ag(111) surface by kinetic Monte Carlo (KMC) method. We carry out KMC growth simulations on surfaces with uniform tensile strain and investigate the surface strain dependence of the island density.

2. Kinetic Monte Carlo Simulation Model

In the KMC method, the epitaxial growth is simulated by sequentially giving rise to each elementary growth process with the event probability [7-10]:

$$\Gamma = \Gamma_0 \exp\left(-E/k_B T\right), \tag{1}$$

where Γ_0 and E are the attempt frequency and the activation energy for each process, k_B is the Boltzmann constant, and T is the substrate temperature. The effective attempt frequency is evaluated by the vibration analyses and the activation energy is obtained from the minimum energy path (MEP) which can be reasonably determined by the nudged elastic band (NEB) method [11]. According to a set of the event probabilities, atoms diffuse thermally on the surface to form islands. In this work, EAM potential developed by Williams *et al* [12, 13] is utilized.

The elementary growth processes we consider here can be classified into six types of KMC events as illustrated in Fig.1:(a)adatom diffusion on terrace, (b)attachment to step, (c)descent from step, (d)diffusion along step, (e)detachment from step, and (f)addimer diffusion on terrace. In the KMC event, the elementary growth processes are distinguished in terms of the adsorption site (fcc or hcp), the initial coordination number n_i, and the final coordination number n_f (ch. Table 1). Moreover, we consider A and B steps inherent in islands grown on the fcc(111) surface: A steps are {111} facets, and B steps are {100}facets, as shown in Fig. 2. For example, the *d* type of event includes 64 elementary growth processes due to the two kinds of step edge, the two kinds of adsorption site, n_i and n_f ranging from 4 to 7. Consequently, total 98 elementary growth processes are considered.

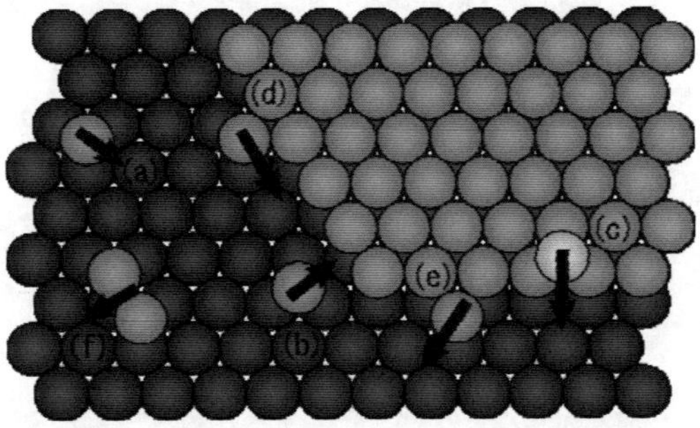

Figure 1. Six types of KMC events.

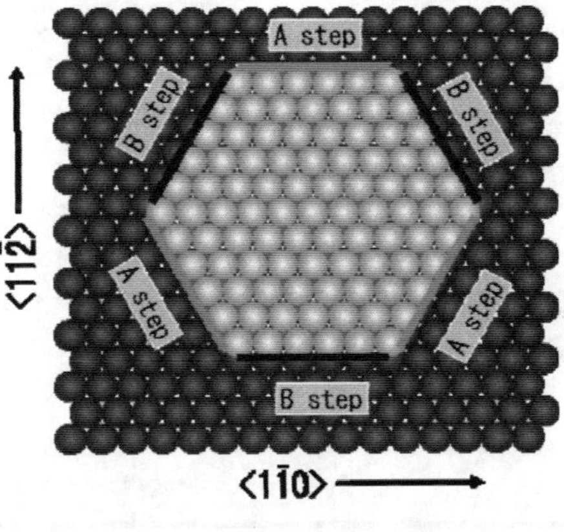

Figure 2. Two steps on fcc(111) surface.

Table 1. Classification of elementary growth processes for each KMC event.

type of event	adsorption site	initial coordination number	final coordination number	step edge
adatom diffusion on terrace	fcc or hcp	3	3	–
attachment to step	fcc or hcp	3	4, 5, or 6	–
descent from step	fcc or hcp	3	4, 5, 6, 7, 8, or 9	–
diffusion along step	fcc or hcp	4, 5, 6, or 7	4, 5, 6, or 7	A or B
detachment from step	fcc or hcp	4, 5, or 6	3	A or B
addimer diffusion on terrace	fcc or hcp	4	4	–

3. Results and Discussion

The surface morphology of epitaxial films is controlled by the substrate temperature, the deposition rate, the biaxial surface strain, and so on. In this study, we investigate influences of the substrate temperature and the surface strain on thin film growth by KMC simulations.

Fig. 3 shows the simulation results of the Ag homo-epitaxial growth on a uniform tensile strained Ag(111) surface. The number of deposited atoms is 54000, which corresponds to 0.06[ML]. The substrate temperature, T, is 100 [K], and the deposition rate, R, is 0.01 [ML/s]. The biaxial surface strains $(\varepsilon = \varepsilon_{11} = \varepsilon_{22})$ are (a) 0, (b) 0.01, and (c) 0.02, respectively. The island shape doesn't depend on the surface strain so much. Fig. 4 shows the island density as a function of the deposition amount in the simulations. For evaluating the island density, we count islands consisting of more than two atoms and neglect adatoms and addimers. As the biaxial surface strain increases, the island density gradually becomes higher as shown in Fig. 4. Thus, the lattice region with the highest tensile strain should be most favorable for nucleation. At the tensile strained region, the lattice spacing is large, and the adatoms are easy to be accommodated.

Next, we investigate influences of the substrate temperature on thin film growth. Fig. 5 shows the simulation results of the Ag homo-epitaxial growth at (a) T=150, (b) T=200, and (c) T=250, respectively. The number of deposited atoms is 54000, R=0.01 [ML/s], and ε =0.02. At T=150 [K], the islands have a dendritic shape with a triangular envelope as shown in Fig. 5(a). At T=200 [K] and T=250 [K], the islands have round and hexagonal shapes, respectively, as shown in Figs. 5(b) and 5(c). At high temperature, the probability of adatom diffusion along step edge is so enhanced that an adatom at the step edge easily attaches to the kink site, and the island has a few kink sites and the shape of the island becomes compact. At low temperature, the probability of adatom diffusion along step edge is so reduced that an adatom at the step edge is associated with another adatom at the step. Then, the new kink site is generated, and the shape of the island becomes dendric. As T increases, the kinetic anisotropy of adatom diffusion along two kinds of step edges is suppressed, so the shape of the islands is more isotropic. On the other hand, the island density is higher at lower temperature, as shown in Fig.6. At high temperature, an adatom can move to the existing island and absorbs at step edge within the deposition interval, and so the formation of new nuclei by isolated adatoms hardly occur.

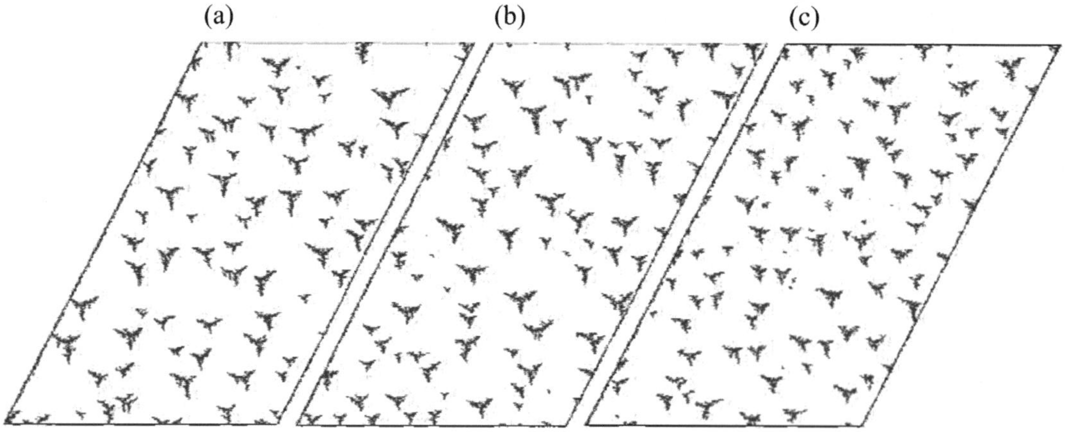

Figure 3. Simulated surface images of self-organized Ag nanoislands on a uniform tensile strained Ag(111) surface with biaxial surface strains of (a)0, (b)0.01, (c)0.02. All simulations are carried out by the deposition amount of 0.06ML (54000 atoms), at T = 100 K and R = 0.01ML/s. The size of the images is 740 × 1230 lattice sites.

doi:10.1088/1742-6596/191/1/012017

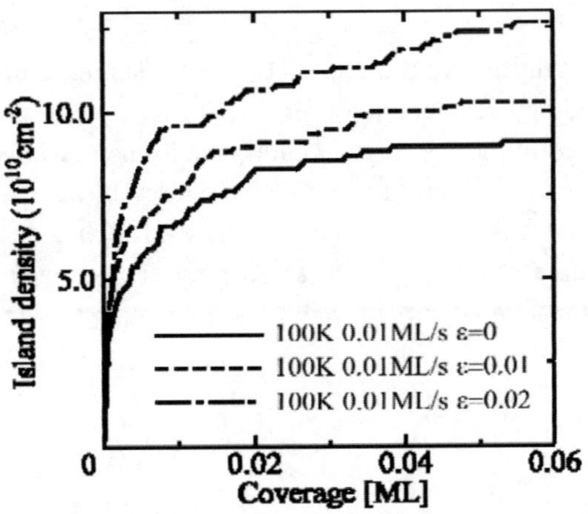

Figure 4. Island density as a function of the deposition amount at T = 100 K and R = 0.01 ML/s.

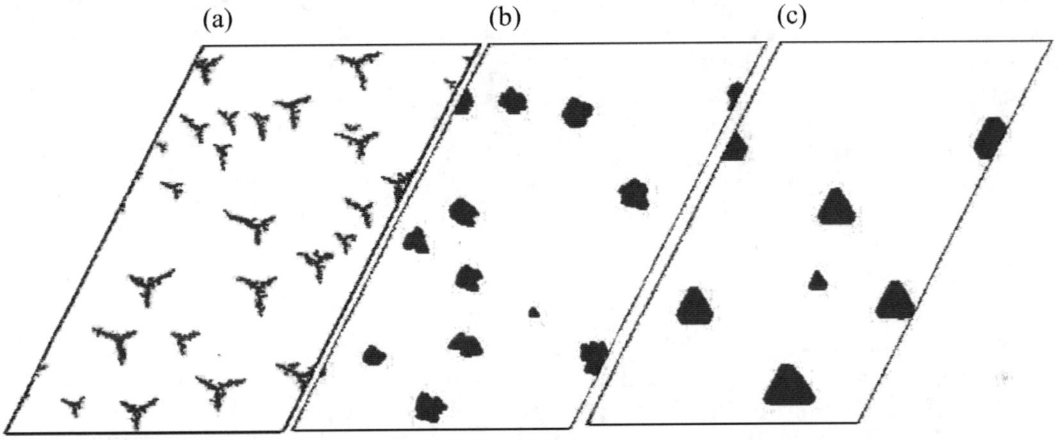

Figure 5. Simulated surface images of self-organized Ag nanoislands on a uniform tensile strained Ag(111) surface at (a) T=150, (b) T=200, (c) T=250 K. All simulations are carried out by the deposition amount of 0.06ML (54000 atoms), at R = 0.01ML/s and ε = 0.02. The size of the images is 740 × 1230 lattice sites.

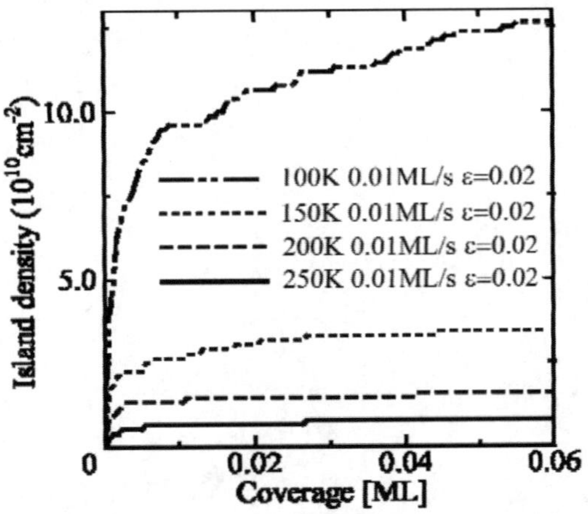

Figure 6. Island density as a function of the deposition amount, at R = 0.01 ML/s and ε = 0.02.

4. Conclusions

We investigated Ag islands grown on uniform strained Ag(111) layer using Kinetic Monte Carlo simulations. We found that the island density depend on the biaxial surface strain and the substrate temperature. As the biaxial surface strain increases, the island density increases. On the other hand, the shape of the island hardly depends on the surface strain. We have also shown that while at low temperature the shape of the island is dendric and the island density is higher, the shape changes to hexagonal and the island density decreases as temperature increases. Thus, the growth behavior on thin film growth, such as the density and shape of island, is tunable by controlling the substrate temperature and the surface strain.

References

[1] Kosiel K 2008 *Vacuum* **82** 951
[2] Jacobsen J, Nielsen L P, Besenbacher F, Stensgaard I, Lagsgaard D, Rasmussen T, Jacobsen K W and Norskov J K 1995 *Phy. Rev. Lett.* **75** 489
[3] Bromann K, Giovannini M, Brune H and Kern K 1999 *Eur. Phys. J.* D **9** 25
[4] Brune H 1998 *Surf. Sci. Rep.* **31** 121
[5] Ling W L, Hamilton J C, Thurmer K, Thayer G E, Figuera J, Hwang R Q, Carter C B, Bartelt N C and McCarty K F 2006 *Surf. Sci.* **600** 1735
[6] Sabiryanov R F, Larsson M I, Cho K J, Nix W D and Clemens B M 2003 *Phys. Rev.* B **67** 125412
[7] Maksym P A 1988 *Semicond. Sci. Technol.* **3** 594
[8] Yamamoto M, Kunizawa K, Fujinami A, Ogata S and Shibutani Y 2007 *Trans. Jpn. Soc. Mech. Eng.* A **73** 490
[9] Yamamoto M, Matsunaka D and Shibutani Y 2008 *Jpn. J. Appl. Phys.* **47** 7986
[10] Kunizawa K, Yamamoto M, Ogata S and Shibutani Y 2008 *J. Soc. Mater. Sci., Japan* **57** 780
[11] Henkelman G B, Uberuaga B P and Jonsson H 2000 *J. Chem. Phys.* **113** 9901
[12] Williams P L, Mishin Y and Hamilton J C 2006 *Modelling Simul. Mater. Sci. Eng.* **14** 817
[13] The potential function for Ag [$V(r)$, $\rho(r)$, and $F(\bar{\rho})$] are available, for retrieval via FTP over the WWW from http://www.ctcms.nist.gov/potentials/Cu-Ag.html.

Third International Symposium on Atomic Technology | IOP Publishing
Journal of Physics: Conference Series **191** (2009) 012018 | doi:10.1088/1742-6596/191/1/012018

Calibration Plate for Digital Holographic Particle Tracking Velocimetry

S Satake[1*], J Taniguchi[1], T Anraku[1], H Kanamori[1], T Kunugi[2], K Sato[3] and T Ito[4]

[1] Department of Applied Electronics, Tokyo University of Science, 2641 Yamazaki, Noda, Chiba, 278-8510, Japan
[2] Department of Nuclear Engineering, Graduate School of Engineering, Kyoto University, Yoshida, Sakyo, Kyoto 606-8501, Japan
[3] Toyota Industries Corp., 2-1 Toyoda cyou, Kariya, Aichi 448-8671, Japan
[4] Japan Science and Technology Agency (JST) and Department of Electronics and Mechanical Engineering, Chiba University, 1-33 Yayoi, Inage-ku, Chiba 263-8522, Japan

E-mail: satake@te.noda.tus.ac.jp

Abstract. A calibration plate for digital holographic particle tracking velocimetry by in-line holography was developed. A quartz plate with a computer-generated hologram pattern is used to determine the distance between an image captured on the plate and a CCD camera to high accuracy. The method allows visual confirmation of the calibration. In addition, the reconstruction of the hologram of a cross marker on plates is demonstrated at distances of 35 mm and 40 mm.

1. Introduction

Holographic particle image velocimetry (PIV) systems using photographic film [1-3] exhibit high quality flow information and allow the recording of instantaneous 3-D velocity fields illuminated by a single beam line. These characteristics are a great advantage over other PIV methods. However the use of photographic film requires considerable reconstruction process time and makes it difficult to capture the time evolution of a particle image from a single frame recording of instantaneous particles dispersed in a flow field. In contrast, digital holographic techniques using a digital camera can easily capture the time evolution of particles. However, the limitation of digital camera resolution means that they are only used with in-line holograms. Digital holography also requires high-speed computer performance to reconstruct particle locations. Previously, a particle tracking velocimetry (PTV) method for macro flow measurements using digital holograms was developed [4]. Our results are in good agreement with theoretical values. In this paper, a method for calibrating digital PTV systems is presented.

Typically, calibration in PIV or PTV is performed using a metal calibration plate with markers for estimation points [5]. When the image of a marker on the plate is captured by a CCD camera, the exact distance between the object (the plate) and the CCD surface of the camera can be obtained. Hence, an image can be calibrated for camera angle and window size. In addition, the calibration of an image can be confirmed conveniently by eye for fluid experiments. However, calibration methods for PIV using digital holograms have not yet been attempted. A calibration technique for PIV using in-line holograms is particularly needed for penetration material owing to the fact that the object beam and reference beam are parallel.

For a holographic image, the image of an estimation point marker is seen only as diffraction fringes. Therefore, since it is not possible to read the image by eye, a calibration plate for digital holographic particle tracking velocimetry with visible real estimation points seen by the camera rather than the fringe image was developed. Using the method presented here, the calibration plate is made of quartz with a calibration hologram pattern drawn on the surface by an electron beam (EB) and chemical etching. The hologram pattern includes estimation points that represent distances determined by a computer-generated hologram. This hologram pattern is then used to determine the exact distance between the object and the CCD camera surface.

2. Digital holography

Figure 1 illustrates the technique of image reconstruction by digital holography. The reconstruction of the maker positions (x,y coordinates) from the fringe images in Fig. 1 is performed by an FFT technique using the Fresnel diffraction equation [6,7]. The FFT technique employs the following relationship:

$$\begin{aligned}\Phi(I,J) &= \int_{-\frac{N1}{2}}^{\frac{N1}{2}}\int_{-\frac{N2}{2}}^{\frac{N2}{2}} H(x,y)G(I,J)\exp[-2\pi i(Ix+Jy)]\mathrm{d}x\mathrm{d}y \\ &= \hat{H}(I,J)G(I,J). \qquad (1)\end{aligned}$$

Here, $\Phi(I,J)$, $H(x,y)$ and $\hat{H}(I,J)$ are the reconstructed image in Fourier space, the hologram image, and the hologram image in Fourier space, respectively. I and J are the pixel points in the x-direction and y-direction. $N1$ and $N2$ are the pixel number in the x-direction and y-direction. $G(I,J)$ is a transform function using the Fresnel approximation, as detailed in Refs. [6,7].

In Eq.(1), $H(x,y)$ is realized in real space. Therefore, the translation from Eq. (1) to get the real maker image $H(x,y)$ is considered. The two-dimensional image of $H(x,y) = \mathcal{F}^{-1}[\Phi(I,J)/G(I,J)]$ is drawn on a material with a penetration. $G(I,J)$ contains the reconstruction information for distance d, the distance between the original object and the hologram image. When the hologram pattern is created by a computer-generated hologram, the distance between the camera and object can be determined highly accurately.

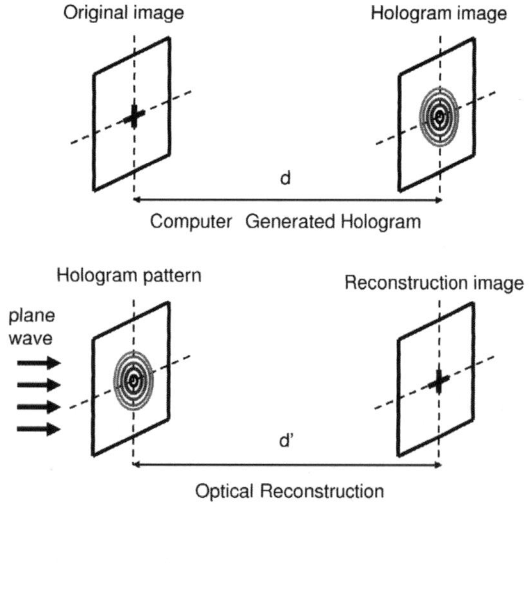

Figure 1. Theory of hologram reconstruction.

3. Fabrication of hologram plate

Figure 2 shows a target marker in the shape of a cross. Figure 3(a) shows the computer-generated hologram of the cross and Fig. 3(b) shows the reconstructed image of the cross obtained from a computer-generated hologram.

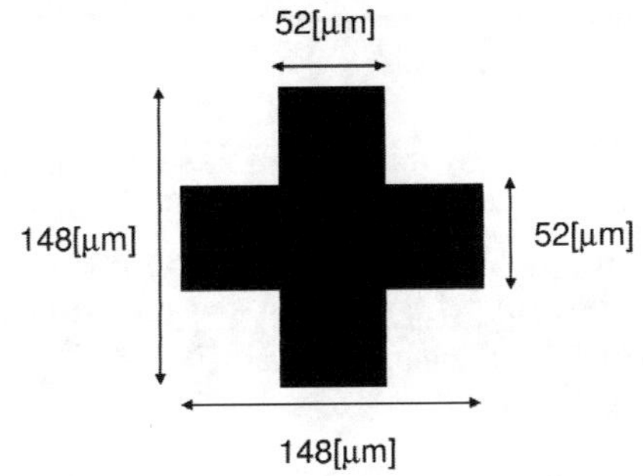

Figure 2. Original cross marker.

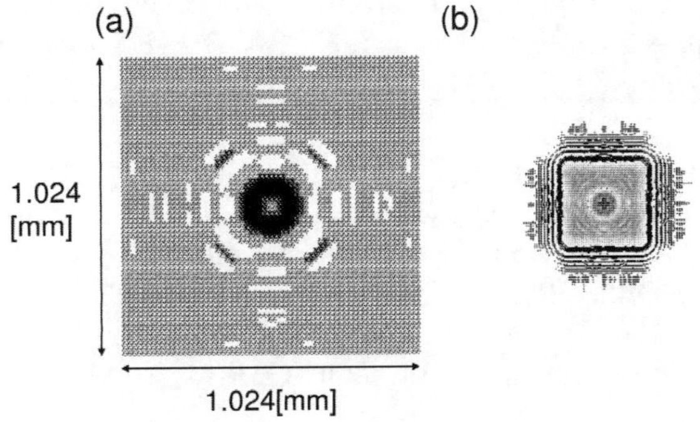

Figure 3. Cross marker for image; (a) Computer-generated hologram of a cross marker (b) Reconstruction of the cross marker from the computer-generated hologram.

The hologram in Fig. 3(a) is reproduced on a quartz plate, as shown in Fig. 4. Quartz is used for the holographic plate, because it is transparent to the laser wavelength of 527 nm used in this study.

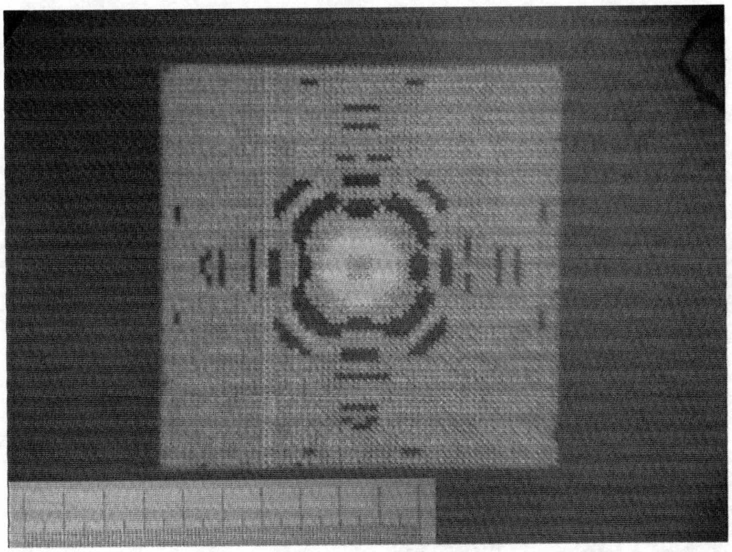

Figure 4. Cross marker from computer-generated hologram is drawn on quartz plate.

The holographic patterns must be precise, and therefore electron beam (EB) lithography is used for their fabrication. However, as quartz is an electrical insulator, charging during the EB writing distorts the resist patterns on the quartz. To prevent charging, a 5-nm conductive polymer ESPACER (Showa Denko K.K.) layer is spin-coated over a 300-nm positive polymethylmethacrylate (PMMA) resist layer.

This sample is exposed using an EB at an acceleration voltage of 10 kV. The exposed resist is developed in a 1:3 mixture of methyl isobutyl ketone (MIBK) and isopropyl alcohol (IPA). A 100-nm thick Cr layer is deposited using an EB evaporation process, and then lifted off in acetone. At the completion of the process the residual Cr patterns form the holographic image.

The computer-generated hologram in Fig. 3(a) was made binary by four gradation steps. It was then numerical reconstructed as shown in Fig. 3(b), confirming that the original image of a cross was restored. A photograph of the image in Fig. 3 (a) transcribed using EB lithography on the quartz plate is shown in Fig. 4. In the photograph the smallest division shown on the scale accompanying the hologram is 10 μm.

4. Experimental setup

Figure 5 shows the experimental setup for reconstructing an image, including a laser, beam expander, traverse facility for holographic plate, and CCD camera. A Nd:YLF laser (Photonic Industries DS20-527, λ= 527 nm) is used, which outputs a pair of laser pulses at a repetition rate of 1 kHz. When the light strikes a marker, it creates hologram fringes that are captured by the high-resolution digital CCD camera (Redlake, MotionPro X3) with a resolution of 1280 pixel $\times$ 1024 pixel (12 μm/pixel).

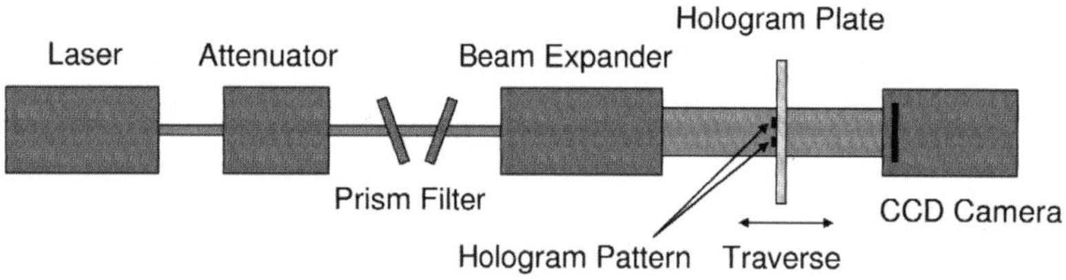

Figure 5. Experimental setup.

5. Result and discussion

In the setup of Fig. 5, the plate can be traversed (moved forward or backward along the optical axis of the setup) by a micro-meter with a resolution of 10 μm. Two plates were designed by the method presented in section 4: plate A (Fig. 6) is designed for focusing at d' =35 mm and plate B (Fig. 7) for focusing at d' = 40 mm.

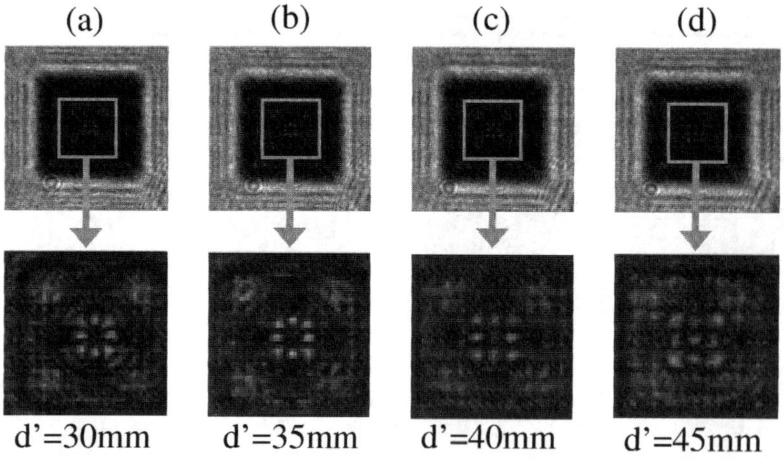

Figure 6. Reconstruction images on quartz plate A in each z-distance.

Figures 6 and 7 show optically reconstructed images of the holograms, along with an enlargement of the center, for the two plates at various distances d' between the plate and

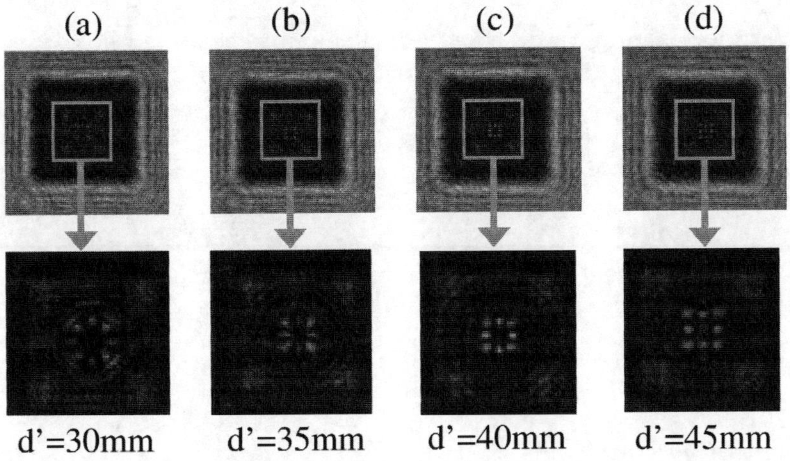

Figure 7. Reconstruction images on quartz plate B in each z-distance.

CCD surface of 30, 35, 40, and 45 mm, recorded by the CCD camera, from a series of images taken between 30 to 45 mm in 1-mm steps.

In Fig. 6, the image of the cross is best resolved in (b) ($d' = 35$ mm). In Fig. 7, (c) ($d' = 40$ mm) is the best resolved. This is in good agreement with the theoretical distances. To more accurately examine the focusing, the brightest points detected for plates A and B are shown in Fig. 8.

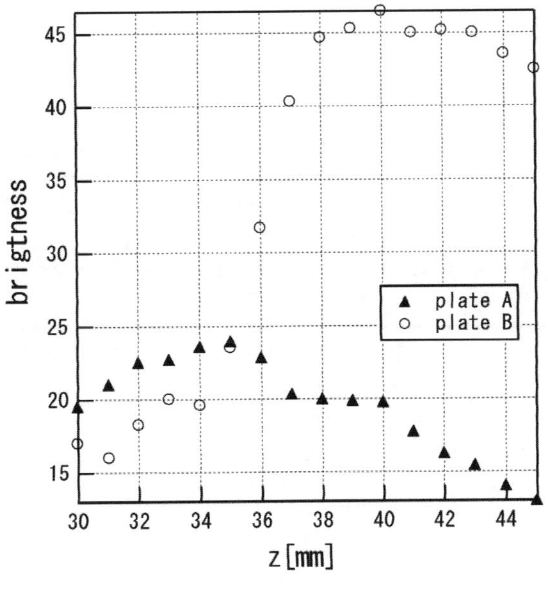

Figure 8. Brightness of cross marker in each z-distance.

The maximum brightness value represents the maximum value measured for 2×2 pixels at four points at the top, bottom, left, and right of the cross. The peak points in Fig. 8 correspond precisely with the theoretical focusing points, namely 35 mm for plate A and 40 mm for plate B.

6. Conclusion

A calibration plate with a pattern for calibration designed by a computer-generated hologram was developed and reconstruction of the hologram pattern at 35 mm and 40 mm was demonstrated.

The object-camera distance in the reconstructions is highly accurate. It is evident that the plate can be the used for calibration for digital holographic particle tracking velocimetry.

Acknowledgments

This study was supported by “ Academic Frontier ” Project for Private Universities: matching fund subsidy from MEXT (Ministry of Education, Culture, Sports, Science and Technology) 2006-2010.The authors thank to Ms. M. Sakai for sample preparation.

References

[1] Barnhart DH, Adrian RJ and Papen GC 1994 *Appl. Opt.* **33** 7159
[2] Meng H and Hussain F 1995 *Appl. Opt.* **34** 1827
[3] Sheng J, Malkiel E and Katz J 2003 *Appl. Opt.* **42** 235
[4] Satake S, Kunugi T, Sato K and Ito T 2004 *Opt. Rev.* **11** 162
[5] Kasagi N and Nishino K 1991 *Exp. Thermal and Fluids Sci.* **4**, 601
[6] Schnars U and Juptner W 1994 *Appl. Opt.* **33** 179
[7] Murata S and Yasuda N 2000 *Opt. Laser Tech.* **32** 567

Vacancy Trapping by Solute Atoms during Quenching in Cu-Based Dilute Alloys Studied by Positron Annihilation Spectroscopy

A Yabuuchi[1], Y Yamamoto[1], J Ohira[1], K Sugita[1], M Mizuno[1,2], H Araki[1,2] and Y Shirai[3]

[1]Division of Materials and Manufacturing Science, Graduate School of Engineering, Osaka University, 2-1 Yamadaoka, Suita, Osaka 565-0871, Japan

[2]Center for Atomic and Molecular Technologies, Graduate School of Engineering, Osaka University, 2-1 Yamadaoka, Suita, Osaka 565-0871, Japan

[3]Department of Materials Science and Engineering, Graduate School of Engineering, Kyoto University, Yoshida-Honmachi, Sakyo-ku, Kyoto 606-8501, Japan

atsushi.yabuuchi@mat.eng.osaka-u.ac.jp

Abstract. Frozen-in vacancies and the recovery have been investigated in some Cu-based dilute alloys by using positron annihilation lifetime spectroscopy. Cu-0.5at%Sb, Cu-0.5at%Sn and Cu-0.5at%In dilute bulk alloys were quenched to ice water from 1223 K. A pure-Cu specimen was also quenched from the same temperature. As a result, no frozen-in vacancies have been detected in as-quenched pure-Cu specimen. On the other hand, as-quenched Cu-0.5at%Sb alloy contained frozen-in thermal equilibrium vacancies with concentration of 3×10^{-5}. Furthermore, these frozen-in vacancies in Cu-0.5at%Sb alloy were stable until 473 K, and began to migrate at 523 K. Finally, the Cu-Sb alloy were recovered to the fully annealed state at 823 K. This thermal stability clearly implies some interaction exists between a vacancy and Sb atom and due to the interaction, thermal equilibrium vacancies are trapped by Sb atoms during quenching.

1. Introduction

Currently, large-scale integration (LSI) Cu interconnects are fabricated by using electroplating techniques. However, accelerated reduction of feature size in Si devices makes it difficult to fill trenches or via holes with copper by using the present techniques. Therefore, a new process has been proposed for fabricating Cu interconnects; filling trenches or via holes with physical vapor deposited copper (PVD-Cu) using high-pressure annealing treatment [1]. What is more, it has already been recognized that the addition of Sb to the PVD-Cu film improves the reflowability.

In the proposed high-pressure reflow process, atomic diffusion is an important factor, and the atomic diffusion rate is proportional to the vacancy concentration. Therefore, it is expected that vacancies are concerned in the improvement of the reflowability. Thus, in this study, we investigated vacancy behaviors in the quenched bulk Cu-Sb dilute alloy and other Cu-based dilute alloys compared with a quenched bulk pure-Cu specimen by using positron annihilation lifetime spectroscopy.

2. Experimental

Cu-0.5at%Sb, Cu-0.5at%Sn and Cu-0.5at%In dilute bulk alloys were prepared by melting 4*N* Cu and 5*N* Sb (4*N* Sn, 6*N* In) in sealed quartz tubes under an argon atmosphere in a furnace at 1373 K and cooled to room temperature. After cooling, these ingots sealed in quartz tubes were heated at 1223 K for 24 h to homogenize the solute distribution. From these ingots, two sheet-like specimens with thickness of about 1 mm were cut out in each alloy. Then, each two sheets of specimen were encapsulated in argon atmosphere quartz tubes again and were annealed at 1073 K for 2 h after that each alloy was maintained at 1223 K for 2h and quenched to ice water. Two sheets of pure-Cu (4*N*) specimen were also water-quenched from the same temperature.

The positron annihilation lifetime was measured for each as-quenched specimen at room temperature. A polyimide film sealed positron source (~1 MBq $^{22}NaCl$) was sandwiched by two sheets of specimen and positron annihilation lifetime spectra were acquired using a fast-fast coincidence positron annihilation lifetime spectrometer. Furthermore, to observe the annealing behavior of frozen-in vacancies, isochronal annealing in steps of 15 min/50 K was performed. Positron annihilation lifetime spectra were acquired for each annealing temperature and all of the measurements were carried out at room temperature. The time resolution of full width at half maximum (FWHM) was 192 ps, and approximately 1×10^6 total counts were accumulated for each lifetime spectrum.

3. Results

Mean positron lifetimes for all specimens before quenching were in a range of 108-111 ps, which are not corresponding to positrons trapped at vacancy-type defects but to positrons delocalized and annihilated in interstitial sites [2-6]. Figure 1 shows the change in mean positron lifetime τ_M for quenched Cu alloys before and during isochronal annealing, the result of as-quenched pure-Cu specimen is also shown. The as-quenched alloys measurement results show that the addition of Sb led to increase the mean positron lifetime. Cu-Sb alloy indicates the highest value of 154 ps and then the value of Cu-Sn alloy is 135 ps. The lifetime of as-quenched Cu-In alloy 112 ps is slightly above the value of Cu perfect crystal 110 ps. Furthermore, mean positron lifetimes of these Cu-Sb, Cu-Sn and Cu-In alloys were fully recovered at 823 K, 573 K and 423 K, respectively. On the other hand, no vacancies were detected in as-quenched pure-Cu specimen with a mean positron lifetime of 110 ps.

To clarify the recovery process, the positron annihilation lifetime spectra of Cu-Sb and Cu-Sn alloys were decomposed into two components. The longer positron lifetime τ_D and the relative intensity I_D of the longer lifetime component are shown in Fig. 2. The dotted horizontal line in the figure shows the calculated positron lifetime which corresponding to positrons trapped at mono-vacancies in a copper crystal. In the case of Cu-Sb alloy, the longer positron lifetime τ_D was stable around 156 ps below 473 K and the relative intensity I_D was also maintained around 90%. Above 523 K, τ_D began to increase and I_D began to decrease drastically. Further, in the case of Cu-Sn alloy, τ_D was stable around 151 ps below 373 K and the relative intensity

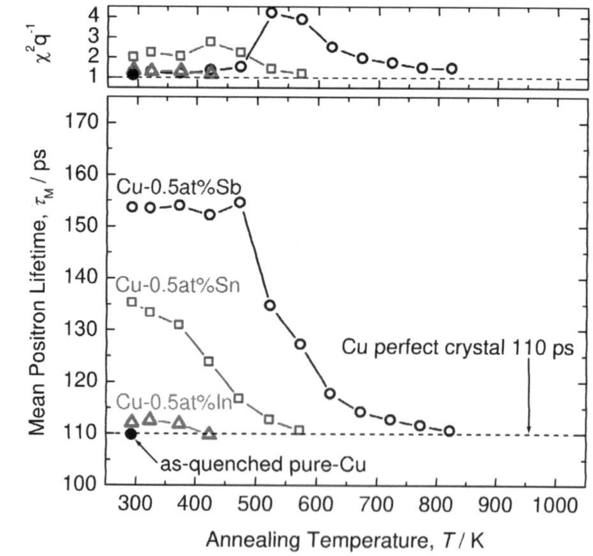

Figure 1. The change in mean positron lifetime τ_M for quenched Cu alloys before and during isochronal annealing. The result of as-quenched pure-Cu specimen is also shown.

I_D was maintained around 74%, and besides, τ_D began to increase and I_D began to decrease above 423 K. In the matter of the Cu-In alloy, we could not resolve a longer positron lifetime τ_D because of the intensity of the longer component was too low.

4. Discussion

In Fig. 1 the value of the $\chi^2 q^{-1}$ is also shown as a measure of the fitting accuracy of the analysis; in case where the quantity of $\chi^2 q^{-1}$ is significantly larger than unity, the measured spectrum is not adequately fitted by the analysis [7,8]. Below 473 K, the $\chi^2 q^{-1}$ value of the Cu-Sb alloy was approximately unity and the mean positron lifetime τ_M was maintained around 154 ps. It indicates that almost all positrons were trapped at mono-vacancy sites and annihilated there. This is also supported by the resolved analysis of the positron annihilation lifetime spectra in Fig. 2. When the Cu-Sb alloy was annealed at 523 K, the $\chi^2 q^{-1}$ value showed over 4 and the mean positron lifetime τ_M decreased significantly which means that the existence of the matrix component besides the defect component because of the decease of the vacancy concentration. This is also indicated in Fig. 2. In the case of the Cu-Sb alloy, above 523 K, the longer positron lifetime τ_D increased and the intensity of the longer component I_D decreased with increasing the annealing temperature, which indicates that frozen-in mono-vacancies began migrating and clustering, and ultimately annihilated at sinks. On the other hand, frozen-in mono-vacancies in the Cu-Sn alloy started clustering at 423 K. It is known that mono-vacancies in pure-Cu can migrate below room temperature [2-6,9,10]. However, the present results show that the addition of 0.5at%Sb led to increase the thermal stability of mono-vacancies in a copper crystal. These results clearly imply vacancies are strongly binding with Sb atoms. A similar phenomenon was observed in Cu-Sn and Cu-In alloys, but the vacancy mobile temperatures were lower compared to the Cu-Sb alloy. From the results, it is reasonable to suppose that the values of the interaction are high in order of Sb-*V*, Sn-*V* and In-*V*, and that interaction causes thermal equilibrium vacancy trapping during quenching.

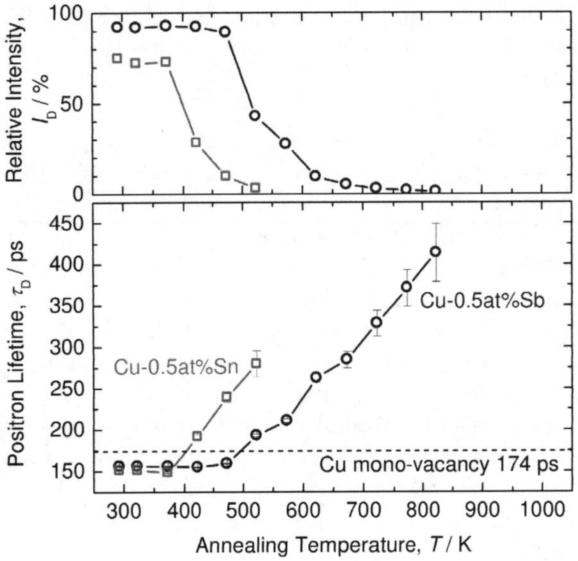

Figure 2. The change in positron lifetime τ_D and the relative intensity of the longer component I_D for quenched Cu-Sb and Cu-Sn alloys before and during isochronal annealing. Error bars are also shown.

Then, in order to estimate the binding energy between a solute atom and a vacancy, we calculated the energy difference between the supercell including a solute atom-*V* pair and that including an isolated solute atom and mono-vacancy using first-principles pseudopotential method as implemented in the Vienna *ab initio* simulation package (VASP) [11,12]. We employed the 2x2x2 supercell containing 32 lattice sites. The calculation results of the binding energies of Sb-*V*, Sn-*V* and In-*V* were 0.46 eV, 0.38 eV and 0.30 eV, respectively. The present experimental results show good agreement with the calculation results. Therefore, the binding energy contributes to trapping thermal equilibrium vacancies during quenching and the difference of frozen-in vacancy concentrations in as-quenched alloys are caused by the value of the binding energy between a vacancy and a solute atom. We estimated vacancy concentrations in as-quenched alloys from the resolved analysis of the positron annihilation lifetime spectra, the value of 1×10^{15} s^{-1} was used as the specific positron trapping rate [13]. As a result, as-quenched Cu-Sb and Cu-Sn alloys contained frozen-in thermal equilibrium vacancies with the concentration of 3×10^{-5} and 7×10^{-6}, respectively. We consider that vacancies are also introduced during deposition in PVD-Cu alloy films fabricated for LSI interconnects by the same mechanism as mentioned above and such vacancies lead to improving the reflowability in the high-pressure annealing process.

5. Conclusions

Frozen-in vacancies in quenched Cu-0.5at%Sb, Cu-0.5at%Sn and Cu-0.5at%In alloys were investigated using positron annihilation lifetime spectroscopy. The results showed that the addition of 0.5at%Sb led to increase the frozen-in vacancy concentration in a copper crystal. In the case of Cu-0.5at%Sn or Cu-0.5at%In alloy, a similar phenomenon was observed but the frozen-in vacancy concentration was lower than that of the Cu-0.5at%Sb alloy. Furthermore, vacancy dissociation temperatures were high in order of Sb-*V*, Sn-*V* and In-*V* pair. These experimental results showed good agreement with the calculation results of the binding energy between a vacancy and a solute atom. In conclusion, the addition of solute atoms which have high binding energy with vacancies lead to promoting thermal equilibrium vacancy trapping during quenching and such frozen-in vacancies promote atomic diffusion. We consider a similar vacancy trapping mechanism occurred during deposition in PVD-Cu films and such frozen-in vacancies might improve the reflowability in the high-pressure annealing process.

Acknowledgement

This study was supported by Priority Assistance for the Formation of Worldwide Renowned Centers of Research – The Global COE Program (Project: Center of Excellence for Advanced Structural and Functional Materials Design) from the Ministry of Education, Culture, Sports, Science and Technology (MEXT), Japan.

References

[1] Onishi T, Fujii H, Yoshikawa T, Munemasa J, Inoue T and Miyagaki A 2003 Thin Solid Films **425** 265
[2] Hinode K, Tanigawa S and Doyama M 1976 J. Phys. Soc. Jpn. **41** 2037
[3] Schaefer H-E 1987 phys. stat. sol. (a) **102** 47
[4] Kluin J-E and Hehenkamp Th 1991 Phys. Rev. B **44** 11597
[5] Staab T E M, Krause-Rehberg R, Vetter B and Kieback B 1999 J. Phys.: Condens. Matter **11** 1757
[6] Eldrup M and Singh B N 2000 J. Nucl. Mater. **276** 269
[7] Kirkegaard P and Eldrup M 1972 Comput. Phys. Commun. **3** 240
[8] Kirkegaard P, Eldrup M, Morgensen O E and Pedersen N 1981 Comput. Phys. Commun. **23** 307
[9] Mantl S and Triftshäuser W 1975 Phys. Rev. Lett. **34** 1554
[10] Mantl S and Triftshäuser W 1978 Phys. Rev. B **17** 1645
[11] Kresse G and Furthmuller J 1996 Comput. Mater. Sci. **6** 15
[12] Kresse G and Furthmuller J 1996 Phys. Rev. B **54** 11169
[13] Eldrup M and Singh B N 1997 J. Nucl. Mater. **251** 132

Metal-insulator Type Nano-granular Magnetic Thin Films -Soft Magnetic Properties, TMR Effect and Their Applications-

S Ohnuma, N Kobayashi, H Fujimori and T Masumoto

Research Institute for Electric and Magnetic Materials,

2-1-1, Yagiyama-minami, Taihaku-ku, Sendai, Miyagi, 982-0807, Japan

Email address: s.ohnuma@denjiken.ne.jp

Abstract. Metal-Insulator type nano-granular magnetic (NGM) thin films have been noteworthily developed as several functional magnetic materials. The formation of nano-granular structure is strongly related to the magnitude of heat formation *(ΔH)* of intergranule materials. Forming a larger *ΔH* allows a clearer granular structure to be obtained. Soft magnetic properties and TMR effect of NGM films are reviewed in this article, including their applications. The HRTEM observation reveals that average distance among neighboring granules mainly controls magnetic properties in NGM films. Co-O based NGM film where granules contact each other has soft magnetism and exhibits an excellent frequency response of permeability, while the film with isolated granules is superparamagnetic state and shows giant magnetostriction effect. Both large electromagnetic noise suppression and low field magnetic sensing effects are demonstrated as their applications.

1. Introduction

The preparation of a soft magnetic film with a nano-granular structure by sputtering method was first reported in 1985[1]. This soft magnetic film, Fe-B-N, was composed of a nano-granular structure of two amorphous phases of Fe-B and B-N, and possessed both large saturation magnetization, *Bs*, (10 kG) and high specific electrical resistivity, ρ, (1000 μΩcm). This study suggested the possibility of developing a new class of materials with good soft magnetic properties in the high frequencies [2]. Since then, extensive studies on several types of nano-granular soft magnetic films Fe-ME-(O, N, F) have been reported in literature [3, 4] (ME=Al, Si, Zr, Hf, Ti, Mg). These films consist of nano-sized Fe granules and an insulative matrix of ME-(O, N, F) with well under 1 nm thickness. We have firstly realized high frequency soft magnetic films with large *Bs*, magnetic anisotropy field, *Hk*, and ρ, by adopting nano-granular Co-ME-O films[5], and also found the films demonstrating a tunneling type magnetoresistance effect (TMR) in the course of the development of Co based soft magnetic films [6]. Magnetic properties of these films change according to the chemical composition, showing a strong correlation with the transport properties. For example, Co-Al-O films for Co concentration below 60 at.% are superparamagnetic, and show a higher ρ than 10^4 μΩcm with giant magnetoresistance (GMR) arising from spin dependent tunneling [6].

The purpose of this report is selectively review the work of metal-insulator type nano-granular magnetic (NGM) films, performed at Research Institute for Electric and Magnetic Materials on the soft magnetic properties and TMR effect. Particular emphasis will be paid on the material design suitable for their applications. A noise suppressor at semi-microwave frequency and low field magnetic sensor are demonstrated as the example of their applications.

2. Preparation and Microstructure

Nano-granular films were prepared by a reactive sputtering under Ar+(O or N) plasma using (Fe, Co)-(M=Al, B, Mg, Si, Zr, RE) alloy targets (RE=Gd, Sm, Y), and by co-sputtering or tandem sputtering under Ar plasma using Fe, Co or alloy target and M-oxide or -fluoride target (M=Al, Mg). The films obtained are in the nano-granular state, consisting of the magnetic nano-granules of Fe, Co or their alloy, and the intergranules of insulating oxide, nitride or fluoride. The intergranular phase of M-oxide in the film was induced by preferential oxidation of M element during deposition. Such a reactive sputtering method using alloy target was useful for controlling accurately the concentration of gas element and for obtaining films with homogenous composition. On the other hand, films showing the TMR effect were mainly prepared on a rotating substrate by tandem sputtering. The key in order to find more suitable M-elements which give good nano-granular structure is the difference of the heat of formation (*ΔH*) between (Fe, Co)-(O, N or F) and M-(O, N or F) compounds. For example, The *Hc* and *Bs* of Co-M-O films decreases and increases monotonically with increasing *ΔH* of the intergranule M-oxide phase, respectively [5,7].

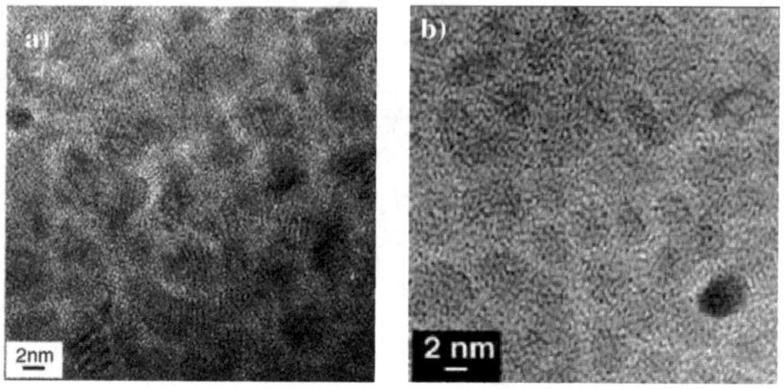

Figure 1. TEM micrograph of nano-granular Co-Al-O films (a: soft magnetic film, b: film with TMR)

To clarify the nano-granular structure, TEM observations were made on Co-ME-O films. Two micrographs of the structure of Co-Al-O films deposited from a $Co_{85}Al_{15}$ and $Co_{75}Al_{25}$ targets are shown in Fig.1. There can be seen a fairly narrow size distribution of granules around several nm. The morphology of Fig. 1(a) shows dark warped granules (magnetic Co) and bright intergranules (Al-oxide). Co granules are in the crystalline fcc/hcp state and coupled ferromagnetically, so that the film has *Bs* of about 1 T and soft magnetic properties due to nano-structural effect [8]. The ρ is large as 300 μΩcm, because of Al-oxide as intergranules, which structure is in the amorphous state. In addition, the film has a large *Hk*. These large *Bs*, ρ and *Hk* give rise to remarkable frequency response of magnetic permeabilities, μ. The experimental results will be presented in the next section 3.

On the other hand, Fig. 1(b) shows that Co granules are well separated and are embedded in Al-oxide matrix. The shape of each granule is likely spherical[6]. Because of this structure, the film becomes superparamagnetic and exhibits large tunnel electric resistance ρ=8x10^4 μΩcm, accompanying with large MR effect. The recent results on the larger TMR materials will be presented in Section 4.

Considerable interest for us is the compositional difference between the granule phase and the intergranules phase in the film. The composition analysis at nano-scale portion in the film examined by means of nm-focused EDX and EELS, using a beam spot diameter of around 1nm[9]. By comparing the results for the granules and intergranule regions, a clear difference between the two compositions was detected. We found the amount of Al element was extremely low in the granules but high in the intergranule region, while the amount of O element was high in the intergranule region and low in the granules. On the other hand, the Co element concentration was fairly high in the granules. These results on the partial composition analysis were in agreement with our previous expectation on the formation of an Al-O glass close to the Al_2O_3 compound as an intergranule phase [5].

3. High frequency soft magnetic properties:

Recent improvements of electromagnetic devices have led to demand for further miniaturization and higher operation of magnetic devices. An important property of magnetic materials required for such applications is frequency response of permeability, which is limited by both eddy current loss and ferromagnetic resonance, *fr*. According to the modified Landau-Lifshitz equation [10] , which predicts the μ-f response, large values of *Bs*, ρ and *Hk* are required simultaneously for excellent high frequency soft magnetic materials. In order to realize a soft magnetic film with larger *Ms, Hk*

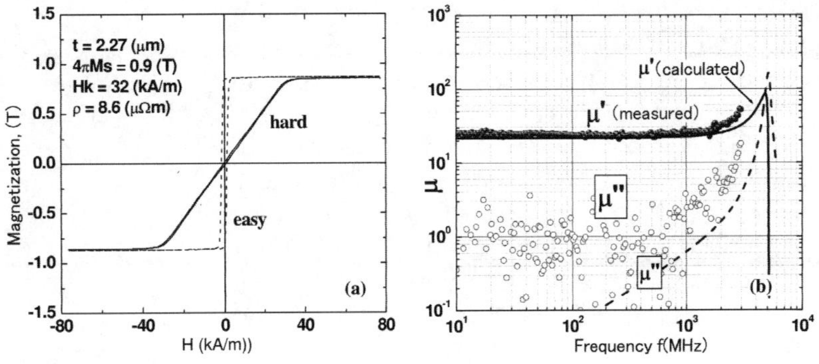

Figure 2. Magnetic properties of soft magnetic Co-Pd-Si-O film (a: magnetization curve, b: μ-f response)

and ρ simultanously, we have paid attention to Co-based films. Our basic idea arises from the fact that generally Co-based alloys exhibit small magneto-striction and high induced magnetic anisotropy. Therefore, if the granule size and the distance between granules (inter-granule region) in the Co-based nano-granular films can be controlled, it may be possible to prepare high frequency soft magnetic films. We have already succeeded for preparing several alloy type of nano-granular soft magnetic films [4]. For example, soft magnetic Co-Zr-O system films with large Hk more than 100 Oe can be obtained in wider compositional range than one of Co-Al-O system films [11]. And we have found that the Ni family elements such as Pt, Pd and Ni certainly improve the properties of the Co-Based nano- granular films, by causing them to be magnetically soft and also enhancing the *Hk* [7]. Fig.2 shows the magnetization curves and μ-f response of Co-Pd-Si-O films, as one of the examples.

Both Increments of *Bs* and ρ in nano-granular soft magnetic (NGSM) film can be readily explained by the formation process of granular structure as mentioned before. However, the great *Hk* value can not be explained by the same reason as *Bs* and ρ. As the excellent frequency characteristics are mainly attributed to the extraordinarily large *Hk* in the Co based NGSM film, our next particular concern is the reason why *Hk*, which is induced by magnetic field [13], is so large. We have carried out systematic investigations on the field induced magnetic anisotropy. Measurements were made of the magnetization curves along and across to the magnetic easy direction and the torque curve[12]. Then, *Hk* and *Ku* were estimated. *Ku* is the uniaxial magnetic anisotropy constant and related to *Hk* as $Ku=(Bs/Hk)/8\pi$. The following properties were found. 1) Torque curve is highly of two-fold symmetry. 2) Easy direction of magnetization is along the field direction applied during the sputtering or annealing. 3) *Ku* can be induced reversibly by the successive re-annealing in field. 4) *Ku* depends on the magnetic elements, being large according to the sequence of Fe based < Co based < Co+Pd based. In order to reveal the origin of *Ku* in NGSM films, the change in *Ku* vs. *x*, where *x*=Co/(Fe+Co) in $(Fe_{1-x}Co_x)_{82}(Al\text{-}O)_{18}$ films were investigated [13]. *Ku* is 10^3 erg/cc in the order of magnitude on the Fe-rich side but it increases with increasing *x*, taking a maximum of about 4×10^4 erg/cc in the Co-rich region. The *Ku* vs. *x* curve is quite different from the relation $Ku=\alpha x^2(1-x)^2$ driven from the model for ferromagnetic binary alloys, in which *Ku* must take maximum just at *x*=0.5, because *Ku* in the model is induced by directional ordering of atom pares [14]. From the above, it is considered that large *Ku* may originate in the nm scale variation of Co base nano-granular structures. In order to clarify this speculation, structural investigations have been performed by means of TEM and small angle X-ray or neutron scattering, now.

The excellent μ-f response, therefore, promise that nano-granular soft magnetic thin films are strong candidate for magnetic cores, such as inductor and magnetic recording head at GHz frequency. $Co_{85}B_{15}$-Pd-O NGSM films, *e.g.*, have ρ around 10^4 μΩcm, *Bs* around 8-9 kG, and *Hk* around 100-150 Oe. The large ρ leads to little eddy current loss, when the film thickness is less than 1 μm, whereas, both large *Bs* and *Hk* result in a considerably high μ'(60-90) and μ''(200-600) near *fr*, and their *fr* is around 3GHz. These data are worthy of not only for the applications using high μ' and quality factor (*Q*), but also for the devices using large μ''. Since $Co_{85}B_{15}$-Pd-O NGSM films have large electrical loss and magnetic loss near *fr, e.g.*, they are expected to be useful as an excellent noise suppressor in semi-micro wave frequency range. A micro strip line, *MSL*, of Zc=50 Ω, whose both ends were connected to network analyzer, was used to evaluate the noise suppression effect [15].

Fig. 3 shows the frequency dependence of noise suppression effect ($\Delta P_{loss}/P_{in}$) up to 6 GHz for several kinds of $Co_{85}B_{15}$-Pd-O films. ΔP_{loss} indicates the real power loss of magnetic films with the size of 15x15 mm, which has been taken by subtracting part by *MSL* from the total power loss [16].

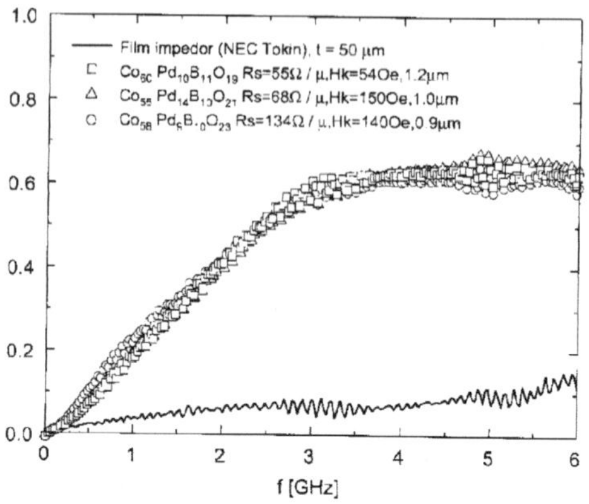

Figure 3. Power losses for Co-B-Pd-O films and a commercial noise suppressor

The result of Film Impedor (NEC-Tokin Co. Japan, 50 μm thickness), which is well known as an electromagnetic noise suppression material at high frequency, is also shown in the figure as a comparison. The $\Delta P_{loss}/P_{in}$ of films increases with increasing frequency up to 3 GHz and keeps constant (around 0.6) in frequencies higher than 3 GHz. It is noted that $\Delta P_{loss}/P_{in}$ of the NGMS films are about 6 times larger than that of the Film Impedor, nevertheless the thickness of Film Impedor is 50 times thicker than those of the NGMS films. An excellent noise suppressing effect in soft magnetic $Co_{85}B_{15}$-Pd-O films is attributed to a large electric loss due to a sheet resistance around 100 Ω, and a magnetic loss or μ'' near fr [17]. We have deposited NGSM film directly on the LSI, which top cover was removed, and found that NGSM film reduced noise level from LSI more than 50 %. In some applications for noise suppression, the thinness of the films is advantageous for heat removal from electric devices.

4. TMR effect:

When the volume fraction of intergranule increases beyond the percolation limit, the electrical conductance obeys a thermal hopping tunnel conductance, and then ρ becomes large. Coincidentally, large TMR appears [6]. The tunnelling conductance has been evident by the temperature (T) variation of ρ according to $\rho = AT^{-2}$, and by the weak temperature dependence of TMR excepting for its anomalous increase at very low temperatures [18]. Further evidences have been obtained by the fact that the current vs. bias voltage curve is parabolic, accompanying Coulomb staircase. The Coulomb staircase implies the existence of the charging effect (Coulomb blockade effect) of Co nanogranules in the tunnelling process.

TMR values obtained in Co-(Al-O) films [6] are in the level of 8 % at room temperature, as shown in Fig.4, which is similar in magnetic field dependence of multilayered Fe/Cr and granular Co-Cu films, characterized by the same negative value irrespective of the longitudinal and transverse cases. We have confirmed that the magnetization curve of Co-Al-O films can be fitted to a Langevin function, implying that these films are superparamagnetic. Nevertheless, in the granular films it is prominent to take useful values of TMR coincident to various ρ values that are larger and smaller than those of multilayers and junctions, respectively. This is quite important merit for applications. We have searched more desirable granular TMR materials. The results are also shown in Fig.5. TMR values up to about 14% can be obtained in the range of $10^4 < \rho < 10^9$ μΩcm. (Fe-Co)-AlF_3 [19] and (Fe-Co)-

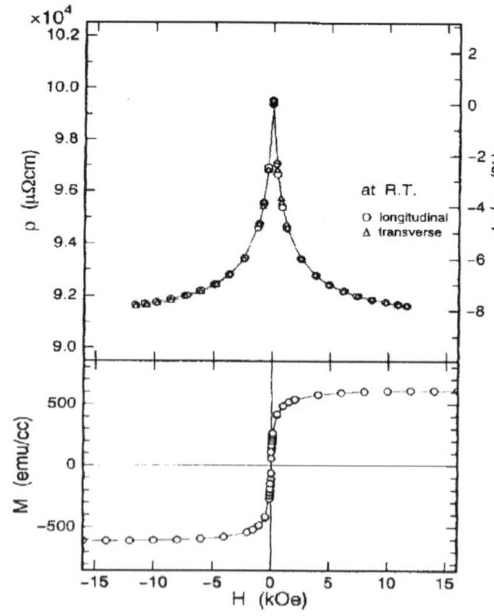

Figure 4. Magnetic field dependence of MR and ρ properties for nano-granular Co-Al-O film

MgF_2 [20] films have larger values of TMR and ρ, compared with Co-(Al-O) films [6]. This may be attributed to the crystalline structure of intergranules, AlF_3 or MgF_2, which is different from the amorphous structure of Al-oxide intergranules.

We have tried to construct a lateral hetero-structure with a soft magnetic film to improve the field sensitivity of TMR, named as GIG sensor [21]. In GIG sensor, it is expected that the large stray field from the soft magnetic film may easily magnetize the nano-granular piece. Fig.6 shows the schematic structure of GIG sensor developed by us. The field response obtained in the GIG devices suitably designed is about 7% of ΔMR per 1 Oe of ΔH with good linearity. The mechanism of the enhancement of the low field response in the GIG sensor is considered as follows: As the soft magnetic film is magnetized in a low external magnetic field, then a large stray magnetic flux from the gap edge of the soft magnetic films passes through the TMR film located in the gap. Then, even at a low external field the TMR film is well magnetized. The magnitude and field dependence of the MR ratio of GIG sensor are considered to depend on the characteristics of both soft magnetic films and TMR materials, as well as GIG structure geometry, such as width and form of the gap and arrangement of the electrodes. The GIG devices are more improved and useful for various kinds of low field sensors, if the structure is optimized. And we are developing the sensor with several companies.

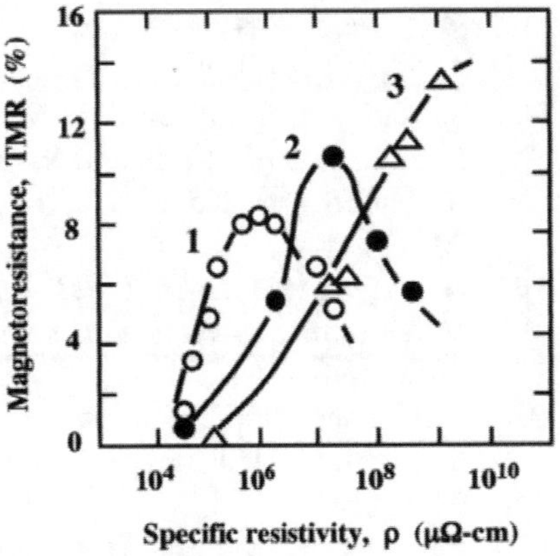

Figure 5. MR vs ρ for various insulating nano-granular films. 1)Co-Al-O, 2)(Fe-Co)-AlF_3, 3)(Fe-Co)-MgF_2

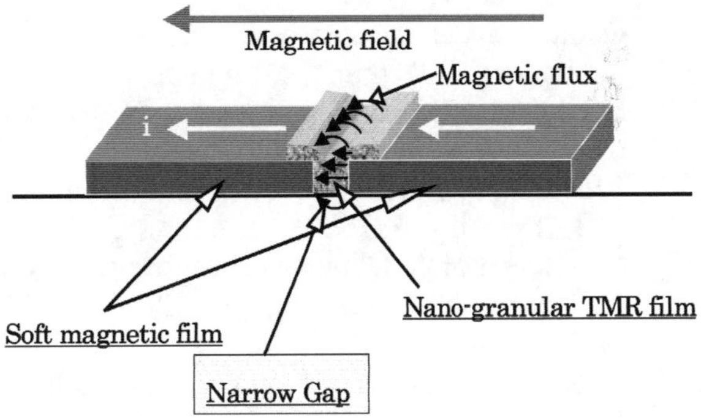

Figure 6. Schematic diagram showing a structure of GIG sensor

5. Closing remarks

Our recent results on nano-granular magnetic thin films are introduced and discussed from the standpoint of high frequency soft magnetic properties and a giant magnetoresistance. We have tried not only high frequency noise suppression effect and low field magnetic sensor, but also several other magnetic devices as their applications. Nano-granular magnetic films also show excellent properties in the areas of magnetic recording media, Hall effect [22], and magneto-optical effect [23]. These remarks are expected to be important for "information technology" in the new generation.

6. Acknowlegement:

This work reported here has been performed in collaboration with M. Ohnuma, H. Mamiya and K. Hono of National Institute for Material Science, and M. Yamaguchi, Y. Shimada and S. Mitani of Tohoku University, and T. Iwasa, H. J. Lee and K. Nagura of Research Institute for Electric and Magnetic Materials. The research was mainly supported by a grant from The Japan Society for the Promotion of Science (JSPS).

7. References

[1] H. Karamon, T. Masumoto and Y. Makino 1985 *J. Appl.Phys.* **57** 3527

[2] H. Karamon 1988 *J. Appl. Phys.* **63** 4306, H. Matsuyama, H. Eguchi and H. Karamon 1990 *J. Appl. Phys.* **67** 5123

[3] A. Makino and Y. Hayakawa 1994 *Materials Science and Engineering* **A181/A182** 1020

[4] S. Ohnuma, H. Fujimori, F. Matsumoto and T. Masumoto 1994 *Materials Science and Engineering* **A182** 892
[5] S. Ohnuma, H. Fujimori, S. Mitani, and T. Masumoto 1996 *J. Appl. Phys.,* **79** 5130
[6] H. Fujimori, S. Mitani and S. Ohnuma 1995 Materials Science and Engineering **B31** 219
[7] S. Ohnuma, N. Kobayashi, T. Masumoto, S. Mitani and H. Fujimori 1999 *J. Magn. Soc. Japan* **23** 23
[8] G. Herzer 1990 *IEEE Trans. Magn.* **26** 1397
[9] M. Ohnuma, K. Hono, H. Onodera, S. Ohnuma, H. Fujimori, and J. S. Pedersen 2000 *J. Appl. Phys.* **85** 817
[10] A. Hosono and Y. Shimada 2000 *J. Magn. Soc. Jpn.* **12** 295
[11] S. Ohnuma, H. J. Lee, N. Kobayashi, H. Fujimori and T. Masumoto 2001 *IEEE Trans. Magn.* **37** 2251
[12] H. J. Lee, S. Ohnuma, H. Fujimori and T. Masumoto 2000 *J. Magn. Soc. Jpn.* **24** 687
[13] S. Ohnuma, N. Kobayashi, and T. Masumoto 1999 *J. Appl. Phys.* **85** 4574
[14] H. Fujimori 1983 *Amorphous Metallic Alloys*, ed F.E. Luborsky (Butterworths Stoneham MA) Chapter 16 pp 300-316
[15] S. Yoshida, H. Ono, S. Ando, F. Tsuda, T. Itoh, Y. Shimada, M. Yamaguchi, K. I. Arai, S. Ohnuma and T. Masumoto 2001 *IEEE Trans. on Mag.* **37** 2401
[16] S. Ohnuma, H. Nagura, H. Fujimori and T. Masumoto 2004 *IEEE Trans. on Mag.* **40** 2712
[17] S. Ohnuma, T. Iwasa, H. Ono, M. Yamaguchi and T. Masumoto 2005 *Diegest of Intermag.* CE-02
[18] S. Mitani, H. Fujimori and S. Ohnuma 1998 *J. Magn. Magn. Mater.* **177-181** 919
[19] N. Kobayashi, *unpublished data*
[20] N. Kobayashi, S. Ohnuma, T. Masumoto and H. Fujimori 2001 *J. Appl. Phys.* **90** 4159
[21] N. Kobayashi, S. Ohnuma, S. Murakami, T. Masumoto, S. Mitani and H. Fujimori 1998 *J. Magn. Magn. Mater.* **188** 30
[22] A. B. Pakhomov, X. Yan and B. Zhao 1995 *Appl. Phys. Lett.* **67**, 3497
[23] E. J. Gan'shina, A. Granovsky, V. Guschin, N. Perov and B. Dieny 1997 *J. Magn. And Magn. Mater.* **165** 320

Effect of annealing on magnetic properties of Nd-Fe-B thin films prepared by ECR ion beam sputtering method

R Tokumaru[1], S Tamano[1], S Goto[1], S Madeswaran[2], K Tokiwa[1,2] and T Watanabe[1]

[1]Department of Applied Electronics, Tokyo University of Science, 2641 Yamazaki, Noda, Chiba 278-8510, Japan

[2]Polyscale Technology Research Center, Tokyo University of Science, 2641 Yamazaki, Noda, Chiba 278-8510, Japan

E-mail: j8107632@ed.noda.tus.ac.jp

Abstract. Nd-Fe-B thin films were prepared by electron cyclotron resonance (ECR) ion beam sputtering and subsequent annealing. The influence of annealing on the magnetic properties and X-ray diffraction patterns of the product films was investigated. Amorphous films deposited at room temperature were annealed at temperatures between 600 and 800 °C. The *c*-axis oriented crystallization of the $Nd_2Fe_{14}B$ phase did not appear by annealing of the buffer layer and magnetic Nd-Fe-B layer deposited at room temperature, and the hysteresis loops of the films indicated magnetic isotropy.

1. Introduction

Recently, Nd-Fe-B permanent magnets have attracted much attention due to their excellent intrinsic properties, such as high saturation magnetization, crystalline magnetic anisotropy, and high coercivity. Thin films with *c*-axis orientation have been studied for their potential applications in microelectronic mechanical systems (MEMS) and magnetic recording media. Preparation of thin films with excellent magnetic properties requires the formation of a well-textured $Nd_2Fe_{14}B$ phase with *c*-axis orientation that is preferentially aligned normal to the film plane, and the grain size of the $Nd_2Fe_{14}B$ phase in the film should be maintained to a single domain size of around 30 nm [1,2]. Two methods are considered for the crystallization of thin films; direct crystallization during film deposition, and crystallization of the as-deposited amorphous films by annealing. The first method has been widely studied, because such films easily acquire magnetic anisotropy perpendicular to the film plane; however, this method may cause the over-growth of grains exceeding a single domain size, which results in degradation of the magnetic properties. On the other hand, grains of single domain size may be obtained using the latter method by control of the film composition and annealing conditions [3]. In this study, the effect of the annealing process on the magnetic properties of Nd-Fe-B thin films was investigated.

2. Experimental procedures

Nd-Fe-B films were prepared using an originally assembled electron cyclotron resonance (ECR) ion beam sputtering system with a 5-cm diameter $Nd_{36.5}Fe_{51.1}B_{12.4}$ alloy target sintered by spark plasma sintering. The background pressure used was below 4.0×10^{-4} Pa and the Ar pressure for sputtering was maintained at 6.0×10^{-3} Pa during film deposition. The following ECR ion beam sputtering conditions were used, acceleration voltage of 1 kV, microwave power output of 100 W, and plasma flux density of 1.2–1.5 mA/cm^2. Amorphous Nd-Fe-B films (100 nm thick) were deposited on Si(100) substrates. W layers were used as buffer (50 nm) and passivation (30 nm) layers. All layers were deposited at room temperature. The as-deposited films were annealed at temperatures between 600 and 800 °C in a tube furnace with Ar gas flow. The O_2 partial pressure was controlled in the range between 10^{-17} and 10^{-19} Pa-O_2 during film annealing using an O_2 partial pressure controller to avoid oxidation of the magnetic layer. The crystal structure was identified using X-ray diffraction (XRD) with Cu $K\alpha$ radiation. The magnetic properties were measured using a vibrating sample magnetometer (VSM) with a maximum field of 20 kOe.

3. Results and Discussion

XRD patterns of the (a) as-deposited film, and films annealed at (b) 600, (c) 700, and (d) 750 °C for 30 min after deposition are presented in Fig. 1. The as-deposited film pattern did not display diffraction peaks assignable to the magnetic layer; only a broad beta-W peak was observed, which implies the magnetic layer is amorphous after deposition at room temperature. The patterns of the post-annealed films did not indicate peaks of the $Nd_2Fe_{14}B$ phase, although diffraction peaks from the W(110), (220) and (211) planes and impurity phases such as Nd-O and a Nd-rich phase, possibly a binary Fe–Nd compound [4], were observed.

Magnified (2θ = 36–44°) XRD patterns of the films annealed at (a) 600 and (b) 700 °C for 30 min are shown in Fig. 2. Weak diffraction lines were confirmed for the film annealed at 700 °C, in contrast to that annealed at 600 °C. However, *c*-axis orientation crystallization of the $Nd_2Fe_{14}B$ phase was not expected, because peaks assignable to the $Nd_2Fe_{14}B(00l)$ phase were not observed.

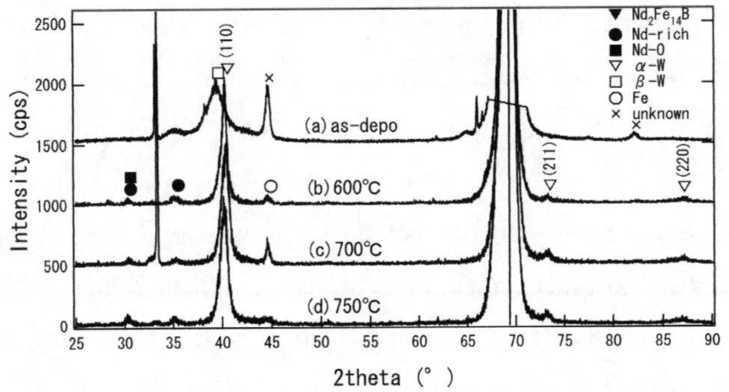

Figure 1. XRD patterns of the (a) as-deposited film and films annealed at (b) 600, (c) 700, and (d) 750 °C for 30 min after deposition at room temperature.

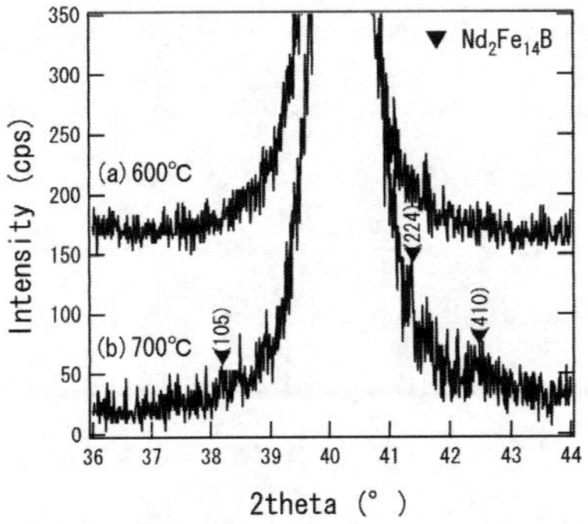

Figure 2. Magnified XRD patterns of the films annealed at (a) 600 and (b) 700 °C.

Magnetic hysteresis loops of the (a) as-deposited film and the films annealed at (b) 600, (c) 700, and (d) 750 °C for 30 min are shown in Fig. 3. Measurements were carried out at room temperature with applied fields perpendicular (⊥) to and parallel (||) to the film plane. No demagnetization correction was carried out for the measurements. The hysteresis loops of the as-deposited film and the film annealed at 600 °C indicate soft magnetic character. The loop of the film annealed at 600 °C indicates higher coercivity than that of the as-deposited film (almost zero). Although no obvious peaks of the Nd-Fe-B phase were observed in the XRD patterns, the coercivity seems to be due to the partial formation of the Nd-Fe-B phase in the film annealed at 600 °C. With increase in the annealing temperature, the hysteresis loops of the films exhibit hard magnetic behavior for the samples annealed between 700–775 °C. Considering the XRD pattern in Fig. 1, this seems to be due to crystallization of the $Nd_2Fe_{14}B$ phase. However, for the films annealed at 700 and 750 °C (Figs. 3(c) and (d), respectively) no obvious difference was observed in the hysteresis loops recorded with magnetic fields parallel and perpendicular to the film plane. This is consistent with the lack of XRD peaks that would indicate the $Nd_2Fe_{14}B$ (00*l*) plane meaning the degree of c-axis orientation. The magnetic properties of coercivity and remanent magnetization were decreased with increase of the annealing temperature over 700 °C. Additionally, large steps are present in the hysteresis loop around the applied field of H=0, measured with both directions of applied field. These steps are due to the magnetic reversal of the soft magnetic phase. It is possible that both the soft and hard magnetic phases are present in the magnetic layer, and that exchange coupling between hard and soft grains is poor.

The respective dependency of coercivity and remanent magnetization on the annealing temperature is shown in Figs. 4 and 5. The sample annealed at 600 °C shows low coercivity, which is reflected by poor crystallization of the hard magnetic phase in the film. The increase of coercivity with further increase of the annealing temperature (650–775 °C) indicates crystalline growth of the magnetic (Nd-Fe-B) phase. The highest coercivity of 8.2 kOe (653 kA/m) was obtained by annealing at 700 °C, with remanent magnetization of 3.4 kG (0.34 T). Annealing of samples above 800 °C resulted in the deterioration of hard magnetic grains, and consequently a non-magnetic phase. The same trend was observed for the remanent magnetization with variation in the annealing temperature, except for the sample annealed at 600 °C, which reveals the presence of the soft magnetic (α-Fe) phase, as observed by XRD.

The results indicate no significant difference in coercivity and remanent magnetization between the parallel and perpendicular directions, even in the samples annealed at 675 and 700 °C, the XRD patterns of which indicated the presence of the Nd-Fe-B phase. This indicates the difficulty of obtaining the *c*-axis orientated $Nd_2Fe_{14}B$ films perpendicular to the film plane using the annealing method. In order to obtain anisotropic *c*-axis textured Nd-Fe-B films, some research groups have applied the film growth technique, which involves deposition at an intermediate temperature between the crystallization between the crystallization temperature and sputtering temperature and sputtering temperature[3,5,6]. Even with a deposition temperature below the temperature required to crystallize Nd-Fe-B phase, thermal assistance during the deposition process may play an important role for the alignment of Nd-Fe-B magnetic grains.

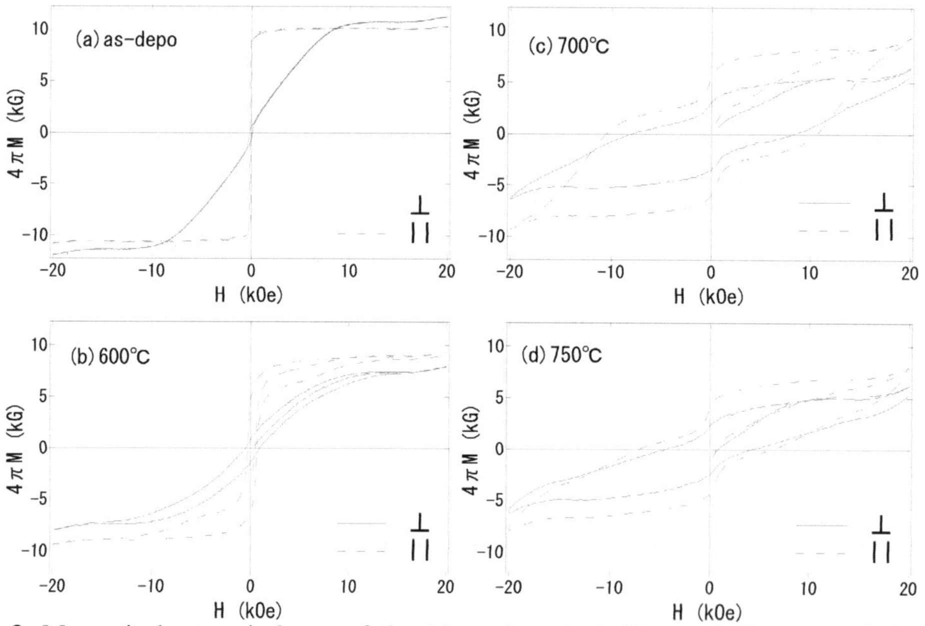

Figure 3. Magnetic hysteresis loops of the (a) as-deposited film, and films annealed at (b) 600, (c) 700, and (d) 750 °C for 30 min after deposition at room temperature. No demagnetization correction was carried out for these measurements.

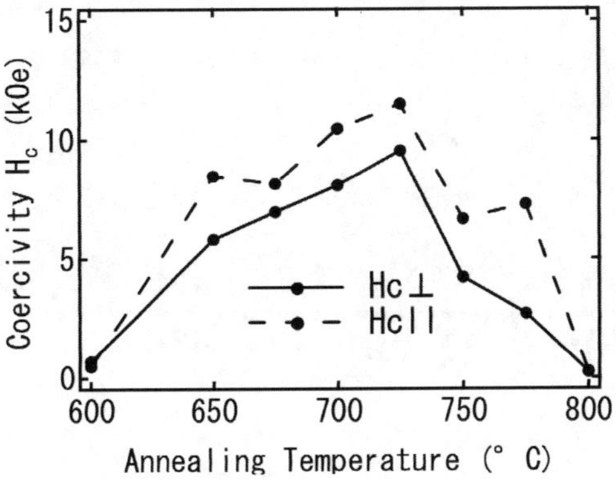

Figure 4. Dependence of the coercivity of Nd-Fe-B thin films on the annealing temperature.

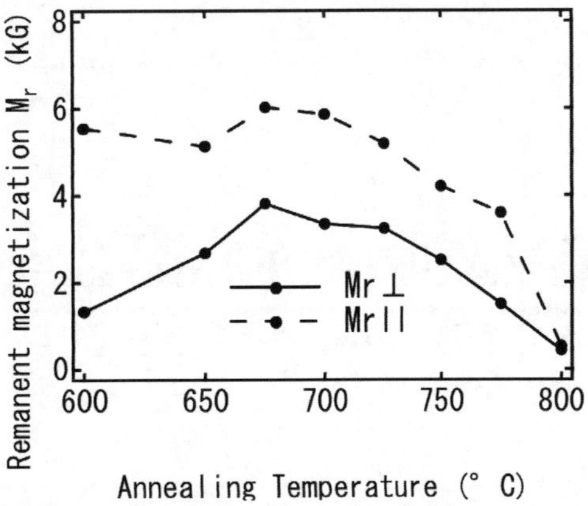

Figure 5. Dependence of the remanent magnetization of Nd-Fe-B thin films on the annealing temperature.

4. Conclusion

Nd-Fe-B thin films were prepared by ECR ion beam sputtering and crystallized by annealing. The influence of annealing temperature on the magnetic properties of the films was investigated. Hard magnetic behavior resulted from annealing temperatures of 650–775 °C. However, the *c*-axis texture of the $Nd_2Fe_{14}B$ phase perpendicular to the film plane was not achieved by crystallization of the buffer layer and magnetic layer deposited at room temperature and subsequent annealing. Hysteresis loops of the prepared films did not indicate perpendicular anisotropic behavior along the *c*-axis. A magnetic Nd-Fe-B film with coercivity of 8.2 kOe (653 kA/m) and remanent magnetization of 3.4 kG (0.34 T) was obtained by annealing at 700 °C.

Acknowledgement

This study was supported by the "Academic Frontier" Project for Private Universities and a matching fund subsidy from the Ministry of Education, Culture, Sports, Science and Technology (MEXT), 2006–2010.

References

[1] Helbest J F 1991 *Rev. Mod. Phys.* **63** 819

[2] Okuda T 2003 *Magn. Soc. Jpn.* **27** 1007

[3] Okuda T, Sugimura A, Eryu O, Serrona L K E B, Adachi N, Sakamoto and Nakanishi A 2003 *Jpn. J. Appl. Phys.* **42** 6859

[4] Chen S.L., W. Liu, C.L. Chen, and Zhang Z.D 2005 *J.Appl.Phys.***98** 033907

[5] Serrona L K E B, Sugimura A, Adachi N, Okuda T, Nakanishi, Motokawa M, Ping D H and Hono K, 2003Appl. Phys. Lett. **82** 1751

[6] Tang S L, Gibbs M R J, Davies H A, Liu Z W, Lane S C, Mateen N E, Du Y W, 2007 *J. Appl. Phys.* **101** 09K501

Journal of Physics: Conference Series **191** (2009) 012022 doi:10.1088/1742-6596/191/1/012022

Optimization of the Magnetic Properties of the $Fe_3B/Nd_2Fe_{14}B$ Bulk Nanocomposite Magnets Prepared by Spark Plasma Sintering

T Fukuzaki[1], K Tanaka[2], K Nishio[2] and R Tamura[1,2]
[1]Polyscale Technology Research Center, Tokyo University of Science
[2]Department of Materials Science and Technology, Tokyo University of Science

E-mail: fukuzaki@rs.noda.tus.ac.jp

Abstract. We have prepared $Fe_3B/Nd_2Fe_{14}B$ bulk nanocomposite magnets at the Nd-poor composition of $Nd_{4.5}Fe_{77}B_{18.5}$ using a spark plasma sintering (SPS) technique, and the influences of the sintering time and the SPS current on the magnetic properties as well as the grain size were investigated. The highest maximum energy product is achieved when the grain size is the smallest on condition that the crystallization of amorphous phase is completed. Excellent magnetic properties i.e., coercivity H_{cj}=248 kA/m, remanence polarization J_r=1.24 T, maximum energy product BH_{max}=81 kJ/m^3 are obtained for the sample prepared with the sintering time of 14.6 sec and the SPS current of 1050A. The microstructure of the optimized magnet shows a fine structure with the average grain size of ~35 nm.

1. Introduction

Nanocomposite Nd-Fe-B bulk magnets are expected as one of the candidates for the next generation permanent magnets since they have a high potential to overcome the magnetic properties of the commercial Nd-Fe-B bonded magnets. Nanocomposite magnets are composed of hard and soft magnetic phases on a nanoscale, and can have large remanence polarization J_r due to an exchange coupling effect between the hard and the soft magnetic phases [1,2]. The strength of the exchange coupling can be controlled by the grain size of the soft magnetic phase and it is expected that strong exchange coupling, i.e., large J_r/J_s, is realized when the grain size is on a nanoscale. The grain size d during grain growth is given by the following equation.

$$d \propto \exp(-Q_D/kT)E_V \Delta t \qquad (1)$$

Here, Q_D and E_v are the activation energy of diffusion and the driving force of crystallization, respectively. From the equation (1), it is understood that the optimization with respect to the heat treatment temperature T and the heat treatment time t is very effective in order to suppress an unfavorable grain growth during sintering. Spark Plasma Sintering (SPS) is known to be one of the powder sintering techniques [3], which is an excellent method for preparation of nanocomposite magnets in a full-density bulk form [4, 5]. For the SPS method, both T and t may be controlled by the SPS current and the sintering time, respectively. Since the highest temperature during sintering can be controlled by the SPS current and the sintering time is roughly proportional to the time that the specimen is exposed to high temperatures, in this study, we have investigated the relationship between the sintering conditions and the magnetic properties of the $Fe_3B/Nd_2Fe_{14}B$ bulk nanocomposite magnets, and have optimized the magnetic properties with respect to the temperature *T* and the time *t*. Furthermore, the microstructure and the average grain size of the optimized magnet has also been investigated in details.

2. Experimental

Precursor alloy ingots of the composition $Nd_{4.5}Fe_{77}B_{18.5}$ were produced by melting pure elements in an arc furnace under an Ar atmosphere. The alloy ingots were then melt-spun onto a Cu wheel rotating with the surface velocity of 50 m/s. The amorphous ribbons were crushed into powder of the size <100 μm and consolidated by the SPS method. During the SPS processes, samples were rapidly sintered with the sintering time of less than 20 sec with the SPS current of 900-1350 A under a compressive stress of 100 MPa. Characterization of the samples was performed by X-ray diffraction, and the microstructure was studied by using a field emission scanning electron microscope (FE-SEM). The magnetic property of the bulk magnets was measured by using a vibration sample magnetometer (VSM).

3. Results and Discussion

Figure 1 presents FE-SEM images of the $Nd_{4.5}Fe_{77}B_{18.5}$ bulk magnets consolidated at the sintering times of (a) 13.2, (b) 14.6 and (c) 16.1 sec with the SPS current of 1050 A. Here, the white area represents the $Nd_2Fe_{14}B$ hard magnetic phase and the dark area the Fe_3B soft magnetic phase. The microstructures are composed of very fine grains of about 30-40 nm and the grain size of the samples increases with increasing sintering time *t*. Figure 2 presents magnetic properties of $Nd_{4.5}Fe_{77}B_{18.5}$ bulk magnets consolidated with the SPS current of 1050A as a function of the sintering time t. It is seen that the coercivity H_{cj} exhibits a maximum at t=14.6 sec: The magnet sintered at 13.2 sec shows a lower H_{cj} value, which is attributed to incomplete crystallization of the $Nd_2Fe_{14}B$ phase due to short sintering time. The H_{cj} value of the sample sintered at 16.1 sec is, on the other hand, slightly higher than that of the sample sintered at 14.6 sec. From figure 1, it is seen that the grain size of the sample sintered at 16.1 sec is larger than that of the sample sintered at 14.6 sec, suggesting that the higher H_{cj} value is due to an excessive grain growth of the Fe_3B and $Nd_2Fe_{14}B$ phases. For the remanence polarization J_r, it decreases monotonously with the sintering time *t*, which is also attributed to the occurence of the grain growth since the J_r/J_s value indicates the strength of the exchange coupling effect between the hard and soft magnetic phases. Thus, the highest BH_{max} is achieved at t=14.6 sec, when the $Nd_2Fe_{14}B$ phases is completely crystallized and the excessive grain growth is suppressed.

Figure 1. FE-SEM images of the $Nd_{4.5}Fe_{77}B_{18.5}$ bulk magnets consolidated at the sintering times of (a) 13.2, (b) 14.6 and (c) 16.1 sec with the SPS current of 1050A.

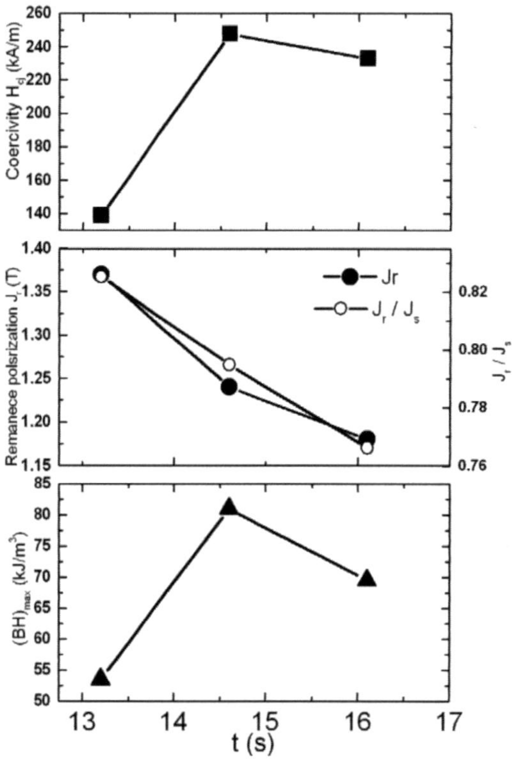

Figure 2. Magnetic properties of $Nd_{4.5}Fe_{77}B_{18.5}$ bulk magnets consolidated at the SPS current of 1050A as a function of the sintering time t.

Figure 3 presents FE-SEM images of the $Nd_{4.5}Fe_{77}B_{18.5}$ bulk magnets consolidated at the SPS current of (a) 900, (b) 1050 and (c) 1350 A with the optimized sintering times. It is seen that the grain size of the samples increases with increasing the SPS current. Figure 4 presents magnetic properties of the $Nd_{4.5}Fe_{77}B_{18.5}$ bulk magnets as a function of the SPS current. Here, the sintering temperature is determined by the SPS current as described in the introduction. The highest magnetic properties are obtained at the SPS current of 1050 A. It is considered that following the above argument, when the SPS current is too small, H_{cj} and J_r become high because of incomplete crystallization of the Fe_3B and $Nd_2Fe_{14}B$ phases, and when the SPS current is too large, H_{cj} and J_r decrease because of an occurrence of the excessive grain growth of the Fe_3B and the $Nd_2Fe_{14}B$ phases. From the optimization with T and t, the sample prepared with the SPS current of 1050A and the sintering time 14.6 sec is found to exhibit the highest magnetic properties i.e., H_{cj}=248 kA/m, J_r=1.24 T, BH_{max}=81 kJ/m^3.

Figure 3. FE-SEM images of the $Nd_{4.5}Fe_{77}B_{18.5}$ bulk magnets consolidated at the SPS current of (a) 900, (b) 1050 and (c) 1350A.

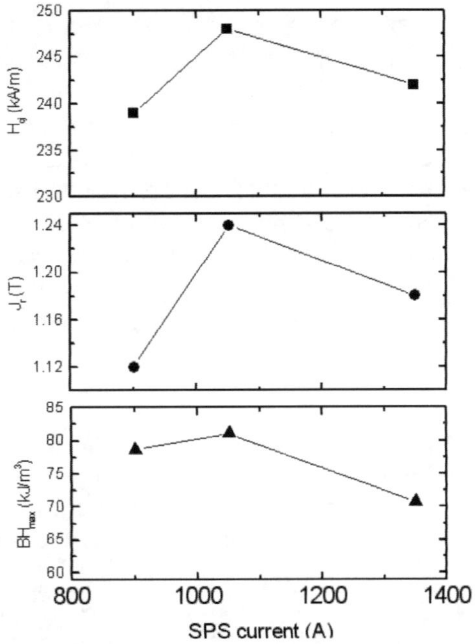

Figure 4. Magnetic properties of the $Nd_{4.5}Fe_{77}B_{18.5}$ bulk magnets as a function of the SPS current.

Figure 5 presents the grain size of the Fe_3B phase as a function of the SPS current. The grain size increases monotonously with increasing the SPS current. As discussed above, the crystallization is not completed when the SPS current is too small, i.e., 900 A, therefore, it is concluded that the SPS current of 1050 A is the optimum current for realizing complete crystallization of the $Nd_2Fe_{14}B$ phase and obtaining fine grains required for efficient exchange interaction. The grain size of the optimized bulk magnet is estimated to be about 35nm.

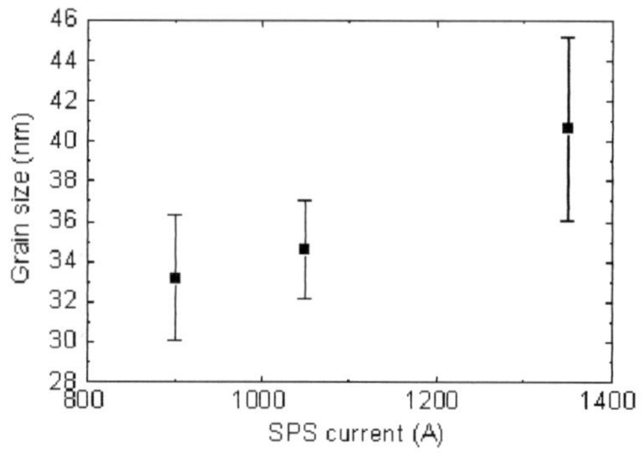

Figure 5. Grain size of the Fe_3B phase as a function of the SPS current.

4. Conclusion

We have investigated the influences of the sintering time and the SPS current in the magnetic properties of $Fe_3B/Nd_2Fe_{14}B$ bulk nanocomposite magnets. An excellent BH_{max} (81 kJ/m^3) is obtained as a consequence of the optimization with respect to the SPS conditions, i.e., when the grain size is the smallest on condition that the amorphous phase is completely crystallized. It has also been demonstrated that the grain size of the nanocomposite magnets can be controlled by the sintering time and SPS current of SPS method, i.e., the optimised sintering time and SPS current are 14.6 sec and 1050A, respectively. The grain size of the optimised bulk magnet is estimated to be about 35nm.

Acknowledgments

This work was supported by "Academic Frontier" Project for Private Universities: Matching Fund Subsidy From MEXT (Ministry of Education, Culture, Sports, Science and Technology), 2006-2010.

A part of this work was conducted in Center for Nano Lithography & Analysis, The University of Tokyo, supported by the Ministry of Education, Culture, Sports, Science and Technology (MEXT), Japan.

Refarences

[1] Coehoorn R, Mooij D B and Waard C 1981 J. Magn. Magn. Matter. **80** 101
[2] Skomski R and Coey J M D 1993 Phys. Rev. B **48** 15812
[3] Tamari N, Tanaka T, Tanaka K, Kondoh I, Kawahara M and Tokita M 1995 J. Ceram. Soc. Jpn. **103** 740
[4] Saito T 2004 J. Matter. Res. **19** 2730
[5] Saito T, Takeuchi T and Kageyama H 2004 IEEE Trans. Magn. **40** 2880

Magnetic properties of glass ceramic wire in Fe_3O_4-TiO_2-SiO_2 system

T Abe[1], T Kishi[1,2] and A Yasumori[1,2]

[1] Department of Materials Science and Technology, Tokyo University of Science,
[2] Polyscale Technology Research Center, Tokyo University of Science
2641 Yamazaki, Noda, 278-8510 Chiba, Japan

yasumori@rs.noda.tus.ac.jp

Abstract. A glassy composite magnetic wire was prepared by elongation during quenching phase-separated melt in two-liquids stable immiscibility region of Fe_3O_4-TiO_2-SiO_2 ternary system. The glass-ceramics in the Fe_3O_4-TiO_2-SiO_2 system has a phase-separation texture, which consisted of Fe_3O_4-TiO_2-rich and SiO_2-rich phases, and shows ferromagnetic properties. The internal textures of wire samples were needle-like structure parallel to the direction of the elongation. The wire sample had larger saturated magnetization and lower coercive force than those of non-elongated bulk one.

1. Introduction

Magnetic materials are widely used for various electronic applications such as a permanent magnet and a magnetic sensor. Recently, preparation of hybrid magnetic material is of particular interest because of the wide range of possible applications: not only in the electronic applications but also in bio- and medical technology. For example, a hyperthermia treatment of tumors using magnetic material is known as one of the treatments for cancer [1], and a ferro-, ferri- or superparamagnetic plate, needle or particles have been investigated [2-7]. An important requirement of the materials for the biomedical application is biocompatibility.

A phase separation in a glass material is very popular phenomenon, and there are two types of phase separation texture, which are called nucleation-growth process and spinodal decomposition one. A texture of the former consists of discrete fine droplets and continuous matrix. A texture of the latter consists of three-dimensionally interconnected two phases. The conventional application of the phase-separated glass is the glass with high thermal and chemical durability and the porous glass in Na_2O-B_2O_3-SiO_2 system that are well known as Pyrex© and Vycor© glass, respectively. Recently, the phase-separated glasses have found broad application such as phosphor [8-12], photocatalytic [13, 14] and magnetic materials [15-18].

We have investigated phase-separated glass-ceramics in Fe_3O_4-TiO_2-SiO_2 system prepared by a melt-quenching method, and the obtained glass-ceramics showed ferromagnetic and electrical conductive properties [17]. The phase-separated textures can be deformed into needle-like shape by elongation during quenching the melt in the system of Fe_3O_4-SiO_2 [15] and CaO-Fe_3O_4-SiO_2 [16], and act as anisotropic functional composite materials [16]. In this study, the melts in the Fe_3O_4-TiO_2-SiO_2

system were quenched with elongation, and the morphology of internal texture, precipitated crystalline phase and magnetic property of the elongated wire and the non-elongated bulk sample were examined.

2. Experimentals

The compositions of two-liquids immiscible melt were $x\mathrm{Fe_3O_4}$-$y\mathrm{TiO_2}$-$(100\text{-}x\text{-}y)\mathrm{SiO_2}$+$5\mathrm{Al_2O_3}$ (mol%, x=15.0, 22.5, 30.0, y=x/3). The small addition of alumina improves the compositional homogeneity of glass melt, and makes the phase separation texture finer in the obtained glass. Starting raw materials of Fe_2O_3, TiO_2, SiO_2 and Al_2O_3 were mixed by ball-milling for 24 hour in ethanol. After drying, the obtained batches were molded into a rod shape by use of a cold isostatic press. The rods were sintered at 1200 °C for 12 h in air. The sintered rod samples were melted for 3 min by use of an infrared imaging furnace (ULBAC, PRC MR-H500) and quenched by turning off the halogen lamps of the furnace. In the preparation method of a wire sample, the tip of the sintered rod was melted in the furnace for 3 min and the melting droplet was contacted with an alumina rod, and then immediately pulled up at 10mm/sec.

The crystalline phases of the quenched samples were analyzed by a powder X-ray diffraction (XRD, SHIMADZU, XRD-6100). Magnetic properties of the samples were evaluated from their magnetization curves, which were obtained by use of a vibrating sample magnetometer (VSM, Riken Denshi Model BHV-55). The melt-quenched samples were cut vertically and polished, and their phase-separation textures were observed by a scanning electron microscope (SEM, JEOL, JSM-5800LV).

3. Results and discussion

3.1. Crystalline phase

The XRD patterns of bulk and wire samples are shown in Figure 1. The sharp diffraction peaks of bulk samples were identified as spinel phase of Fe_3O_4 or Fe_3O_4-TiO_2 solid solution. An amorphous halo due to SiO_2-rich phase was slightly observed at around 22° in all the samples. In the bulk sample of x = 30.0, the weak diffraction peaks of hematite (α-Fe_2O_3) phase were also observed. The precipitation of α-Fe_2O_3 phase is principally due to oxidation of iron (II) to iron (III). The reason for the oxidation of iron ion in the sample of x = 30.0 is that this sample has less amount of SiO_2, which prevents oxygen from diffusing into the Fe_3O_4-TiO_2 rich phase, than the other samples.

The peaks of all the wire samples contained both spinel and α-Fe_2O_3 phases. The wire sample is easy to oxidize compared with the bulk one, because the wire sample has larger specific surface area than bulk one. The relative peak intensity of α-Fe_2O_3 phase in x = 30.0 wire sample is obviously higher than the other ones, indicating that the more amount of iron ion in this sample is oxidized than that in the other ones. This is due to less amount of SiO_2 same as the bulk samples.

3.2. Magnetic properties

The magnetization curves of the bulk and wire samples are shown in Figure 2. Table 1 summarizes the results of the present magnetization measurements: magnetization at maximum applied magnetic field: 15 kOe, M_S, residual magnetization, M_R and coercive force, H_c. The M_S of both bulk and wire samples increased with increase of Fe-Ti content, x. Since the M_S of sample is not proportional to the amount of x, this is due not only to the increase of the amount of spinel phase precipitated in the Fe_3O_4-TiO_2-rich phase, but also to strengthening exchange interaction between iron ions. Their H_c decreased with increasing x. This decrease of H_c can be considered to arise from some difference of magnetic domain structure in the samples.

In the samples of x = 15.0 and 22.5, the M_S and M_R of the wire samples are higher than those of bulk samples. This difference of magnetic properties is most likely due to difference of cooling rate between the bulk and the wire samples resulting in change of the amount of precipitated spinel phase. On the other hand, the M_S of the x = 30.0 wire sample lower than that of bulk one, because much larger amount of spinel phase changes to α-Fe_2O_3.

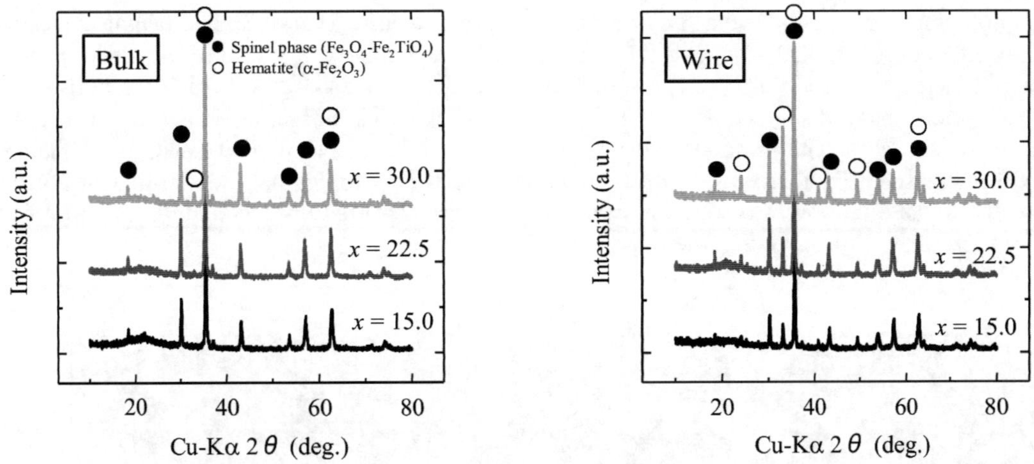

Figure 1. The XRD patterns of the melt-quenched bulk and wire samples with the composition of xFe$_3$O$_4$-yTiO$_2$-(100-x-y)SiO$_2$+5Al$_2$O$_3$ [mol%, $y=x/3$].

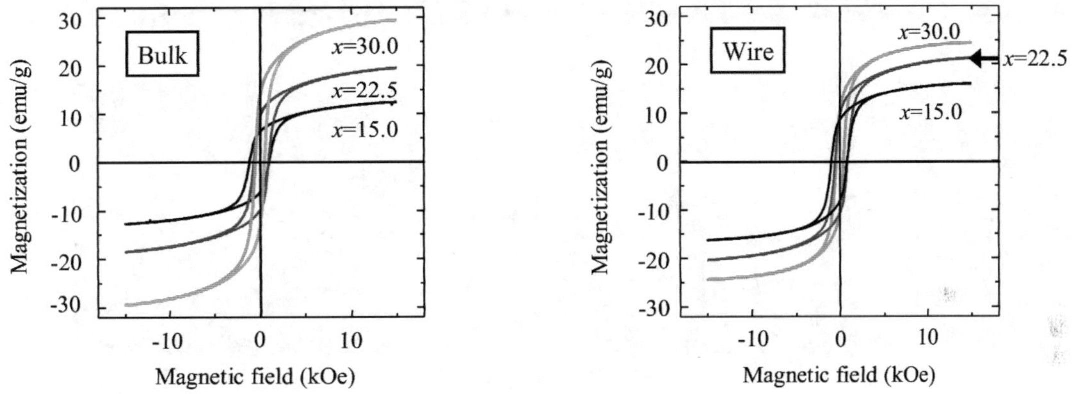

Figure 2. The magnetization curves of the melt-quenched bulk and wire samples with the composition of xFe$_3$O$_4$-yTiO$_2$-(100-x-y)SiO$_2$+5Al$_2$O$_3$ [mol%, $y=x/3$].

Table 1. Magnetization at maximum applied magnetic field: 15 kOe, M_S, residual magnetization, M_R and coercive force, H_c, of the bulk and wire samples with the composition of xFe$_3$O$_4$-yTiO$_2$-(100-x-y)SiO$_2$+5Al$_2$O$_3$ [mol%, $y=x/3$].

x mol%	Sample shape	M_S [emu/g]	M_R [emu/g]	H_c [kOe]
15.0	Bulk	12.43	6.355	1151
	Wire	16.01	8.488	983.7
22.5	Bulk	18.98	9.697	1151
	Wire	20.82	10.13	557.4
30.0	Bulk	29.40	13.49	528.3
	Wire	24.81	10.44	435.9

3.3. Internal texture

The SEM photographs of the bulk and the wire samples are shown in Figure 3. The bright and dark areas are attributed to Fe_3O_4-TiO_2-rich and SiO_2-rich phases, respectively, because of the difference of the emission efficiency of secondary electrons. The microstructures in the bulk samples consist of grains of SiO_2-rich phase dispersed in a matrix of Fe_3O_4-TiO_2-rich phase, which was formed by nucleation-growth process. The grain size of SiO_2-rich phase and the area of Fe_3O_4-TiO_2-rich matrix increase with increase of Fe-Ti content. On the contrary, the wire samples had needle-like SiO_2-rich phases oriented in the direction of elongation in parallel. Such a distorted phase-separation texture has a possibility to introduce anisotropy in the glass-ceramics [16].

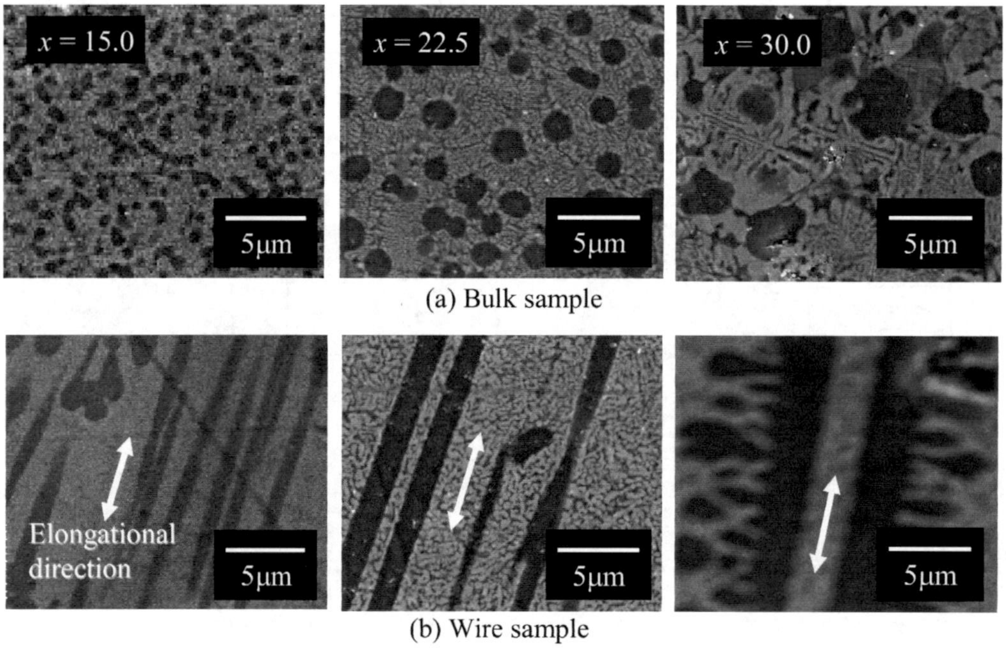

Figure 3. SEM photographs of the melt-quenched bulk and wire samples with the composition of xFe$_3$O$_4$-yTiO$_2$-(100-x-y)SiO$_2$+5Al$_2$O$_3$ [mol%, $y=x/3$].

4. Summary

The glassy composite magnetic material with needle-like phase-separation texture was obtained by quenching the two-immiscible phase-separated melt in the Fe_3O_4-TiO_2-SiO_2 ternary system. The precipitated crystalline phase was mainly the spinel phase of Fe_3O_4 or the Fe_3O_4-TiO_2 solid solution. In the sample with lower SiO_2 concentration, the larger amount of α-Fe_2O_3 phase was precipitated. The texture formed by nucleation-growth process that consisted of the Fe_3O_4-TiO_2-rich and SiO_2-rich phases was observed at the bulk samples, and the SiO_2-rich phases in the wire samples were elongated. The direction of the elongated texture was parallel to that of elongation of the melt. All the samples showed ferromagnetic properties on their magnetization curves. The saturated magnetization increased and the coercive force decreased with increase of Fe-Ti content. It is thought that large amount of spinel phase was precipitated and that the magnetic domain structure changed from single magnetic domain to multiple one with increase of Fe-Ti content in the sample.

Acknowledgements
This work was supported by "Academic Frontier" project for private universities: matching fund subsidy from MEXT (Ministry of Education, Culture, Sports, Science and Technology), 2006-2010.

References
[1] Atsumi T, Jeyadevan B, Sato Y and Tohji K 2006 *J. Magn. Soc. Jpn.* **30** 555
[2] Kong G, Braun R D and Dewhirst M W 2000 *Cancer Research* **60** 4440
[3] Ohura K, Ikenaga M, Nakamura T, Yamamuro T, Ebisawa Y, Kokubo T, Kotoura Y and Oka M 1991 *J. Appl. Biomater.* **2** 153
[4] Xiang S –H and Liu J 2008 *Int. J. Heat & Mass Transfer* **51** 5486
[5] Arcos B, Real R P del and Vallet-Regi M 2002 *Biomaterials* **23** 2151
[6] Shimizu T, Asano H, and Matsui M 2007 *J. Magn. Magn. Mater.* **310** 1835
[7] Bahadur D, Giri J 2003 *Sadhana* **28** 639
[8] Yao Z, Ding Y, Nanba T and Miura Y 1998 *Mater. Sci. Res. Int.* **4** 141
[9] Yao Z, Ding Y, Nanba T and Miura Y 1998 *Mater. Sci. Res. Int.* **4** 244
[10] Ohgaki T, Higashida A, Soga K and Yasumori A 2007 *J. Electrochem. Soc.* **154** J163
[11] Ohgaki T, Nagumo Y, Soga K and Yasumori A 2007 *J. Ceram. Soc. Jpn.* **115** 201
[12] Hijiya H, Kishi T and Yasumori A 2008 *J. Ceram. Soc. Jpn.* **116** 1255
[13] Sone Y, Nishio K and Yasumori A 2004 *Mater. Res. Soc. Symp. Proc.* **849** KK3.13
[14] Sone Y, Ohgaki T, Nishio K, Yamashita T and Yasumori A 2006 *J. Photopolym. Sci. Tech.* **19** 111
[15] Yasumori A, Koike A, Kameshima Y, Okada K, Yano T, Yamane M and Inoue S 2000 *J. Ceram. Soc. Jpn.* **108** 813
[16] Yasumori A, Koike A, Kameshima Y, Okada K and Nishio H 2002 *J. Non-Cryst. Sol.* **297** 239
[17] Kono K, Kishi T, Ohgaki T and Yasumori A 2008 *J. Phys: Conference Series*, **106** 012010
[18] Nakamura H, Kishi T, Ohgaki T and Yasumori A 2008 *J. Phys: Conference Series*, **106** 012009

Dependence of deposition parameters and layer thickness on the characteristics of Nd-Fe-B thin films

S Madeswaran[1], R Tokumaru[2], S Tamano[2], S Goto[2], K Tokiwa[1,2] and T Watanabe[1]

[1]Polyscale Technology Research Center, Tokyo University of Science
2641 Yamazaki, Noda, Chiba 278-8510, Japan

[2]Department of Applied Electronics, Tokyo University of Science, Japan
2641 Yamazaki, Noda, Chiba, 278-8510, Japan

E-mail: smades@rs.noda.tus.ac.jp

Abstract. Textured Nd-Fe-B thin films with hard magnetic properties were prepared on a Ta (110) buffer layered glass substrates using radio frequency (RF) sputtering deposition. We investigated the influence of substrate temperature, sputtering gas pressure, RF power and film thickness on their microstructural and magnetic properties. Composition in the Nd-Fe-B thin films prepared using the same target with an Nd/Fe ratio of 0.32 was markedly changed (varied between 0.21 and 0.31) depending on the Ar pressure and the RF power. Well-textured Nd-Fe-B films grown at a deposition pressure of 7.0 Pa, a temperature of 550 °C, and a power of 100 W revealed better magnetic properties: J_r=1.1 T, H_c = 1130 kA/m and $BH_{(max)}$ = 236 kJ/m^3.

1. Introduction

Recent investigations on the hard magnetic Nd-Fe-B thin films with easy magnetization axis have opened up great possibilities of applications like micro-electro-mechanical devices [1] and data recording media [2]. The Nd-Fe-B magnets have also extended for high power actuators and susceptible to high temperature irreversible demagnetization [3] owing to their high energy product value. The high energy product value depends on high remanence and high coercivity that always enhances the performance of a device [4]. The remanence values are greatly influenced by the grains alignment which is considerably attained by the texturing of the sample [5, 6]. The coercivity has at least half the value of remanence for a high performance film. The intrinsic properties of the magnetic phase and the direct exchange or long-range magnetostatic interactions between the grains of phases are the deciding factors of coercivity of the magnetic materials. Microstructure of a magnetic material plays a key role in determining the interaction mechanism. Therefore, the study on microstructure is not only essential for understating the magnetic phenomena but for tailoring the magnetic properties.

In general, the strong c-axis orientation and enhanced perpendicular magnetic properties in the crystallized films are potentially modified by the deposition process [7]. Several works were reported on the effects of deposition parameters on Nd-Fe-B films properties [8-10]. However, in these reports, the film properties seemingly varied very much depending on the deposition methods used for film preparation apart from the deposition parameters. The suitable conditions for making aligned films are more likely depending on the sputtering system besides the substrate material, target composition, substrate temperature, sputtering gas atmosphere, and deposition rate [6, 8-11]. In this report the effects of various process parameters on the magnetic properties of Nd-Fe-B films prepared by RF sputtering are investigated.

2. Experiment

The films were deposited in a high vacuum chamber with a base pressure of ~10^{-5} Pa by sputtering a Nd-rich target with a nominal composition of $Nd_{20}Fe_{64}B_{16}$ prepared by Spark Plasma Sintering. Ta buffer layer with the thickness of 50 nm was deposited onto the glass substrate prior to depositing the Nd-Fe-B film, to enable texturing of $Nd_2Fe_{14}B$ phase and also to prevent diffusion of the substrate materials into the film [12]. A cap layer of 20 nm thick Ta was also deposited over the Nd-Fe-B film to product it from oxidation on contact with air. A resistive heater was used to regulate the substrate heating. The films were grown at substrate temperatures (T_s) of 300, 400 and 500 ˚C and the sputter power varied between 50 and 150 W. The distance between target and substrate was fixed at 60 mm. The sputtering Ar gas pressure was changed from 1 to 17 Pa. A sputter rate of 20 nm /min at 100 W was used to study the effect of film thickness. While we adjusted one deposition parameter, all others were remaining unchanged. X-ray analysis was performed with a Rigaku Ultima-III diffractometer utilizing Cu-Ka radiation. Microstructural investigations were performed with a Hitachi S-4200 digital scanning electron microscope (SEM), equipped with energy dispersive X-ray (EDX) detector. The film thickness was measured by cross-sectional SEM. Magnetic properties were measured by Vibrating Sample Magnetometer (VSM) with a maximum field of 20 kOe. The magnetic properties were measured both perpendicular and parallel direction to the film plane.

3. Results and Discussion

3.1. Effect of Temperature

The XRD patterns of as-deposited films [layer form-Ta(50nm)/NdFeB(300nm)/Ta(20nm)] grown at temperature between 300 and 500 ˚C on a Ta-buffered glass substrate at a power of 100W and a Ar pressure of 7 Pa are shown in figure 1. For the film deposited at temperature 300 ˚C, peaks of $Nd_2Fe_{14}B$ crystalline phase are absent, however, Ta (110) and a peak at $2\theta = 33.5^\circ$ that may corresponding Fe_3B phase (JCPDS data 34-1198) appear, indicating that the Nd-Fe-B film is amorphous. When T_s increases from 300 to 400 ˚C, another peak emerges at 44.6° which can be attributed to (006) of $Nd_2Fe_{14}B$ or α-Fe phase. In the case of film deposited at 500 ˚C, high intensity peaks corresponding (004), (006) and (008) plane of crystalline $Nd_2Fe_{14}B$ phase occur with a strong (00*l)* texture, tending to be perpendicular to the plane of the film. Additionally, the x-ray pattern also shows that there are some inclusions of Nd-rich and Nd_xO_y phases. Based on the reports by Minowa et al (13) and Yue et al (14), it can be say that the Nd-rich phase consists of complex Nd containing intermetallic compound and Nd_xO_y. Note that regardless of different deposition temperatures, all these films show the presence of Fe_3B phase.

Figure 2 shows the XRD patterns for the films deposited at 300, 400, and 500 ˚C after being annealed at 600 ˚C for 15 min. Of the three films, the film deposited at 500 ˚C exhibited notably high intensity peaks of the (004), (006) and (008) plane which are the most intense in the tetragonal $Nd_2Fe_{14}B$ phase with the preferred *c*-axis orientation perpendicular to the plane of the film. This suggests that annealing treatment eventually enhances the crystallization in the film grown at a higher deposition temperature than the films grown relatively at lower temperatures. This is more evident in case of the 004 peak with increased sharpness. It is interesting to note that the peak appeared at 33.5° in the as-deposited state completely disappeared after annealing, suggesting that it might be related to the unstable Fe_3B phase in these films.

The magnetic behavior measured by VSM (figures not shown) showed soft magnetic phases for as- deposited films at 300 and 400˚C whereas the film deposited at 500 ˚C revealed hard magnetic phase. This is consistent with x-ray diffraction data displayed in figure 1. The hysteresis loops measured at room temperature for the films annealed at 600 ˚C are shown in figure 3. The solid and dashed lines represent measuring field direction along perpendicular (out-plane-⊥) and parallel (in-plane-||) to the film plane, respectively. No demagnetization correction was performed for any of the measurement. The loops indicate that the film grown at 500 °C and annealed at 600 °C demonstrated better magnetic properties when compared to the ones grown at lower temperatures after annealing at the same temperature. It can also be seen that the ratio of $J_{r\perp} / J_{r\parallel}$ for the film deposited at 500 °C is higher (3.2) than the films deposited at 300 and 400°C (2.2 and 2.5 respectively). Being consistent with the XRD results, the magnetic properties also confirm that better crystallization of magnetic $Nd_2Fe_{14}B$ phase occurs only for direct deposition at high temperature while it is less for the films produced at low temperatures and annealed at high temperature. Tang *et al* [7] observed constant out-plane coercivity, H_c, for both low and high temperature deposition with subsequent annealing at high temperature. This is mainly because of different target composition, i.e, high Nd content in our case, which leads to the parallel growth of Nd-rich and Nd_xO_y phases. It is expected that the Nd_xO_y phase might support the growth of $Nd_2Fe_{14}B$ phase with c-axis texture when the deposition temperature is high, however, it hinders growth of the texturing $Nd_2Fe_{14}B$ phase at low deposition temperatures and subsequent annealing treatment [11, 15].

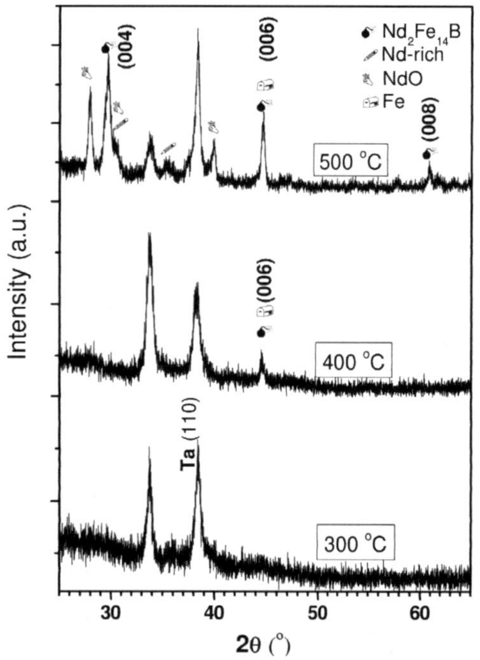

Figure 1. XRD patterns for the films deposited at different substrate temperature.

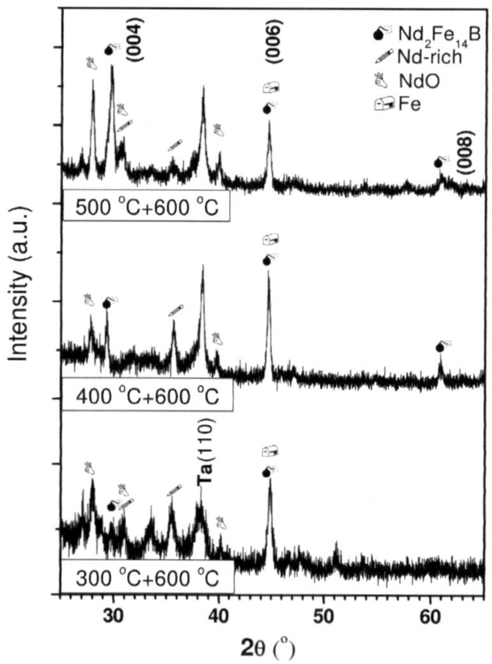

Figure 2. XRD patterns for the films deposited at different substrate temperature and annealed at 600 °C.

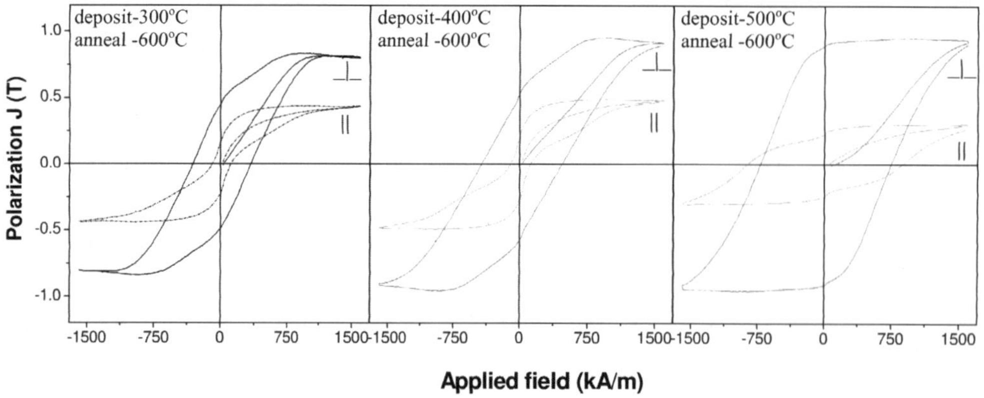

Figure 3. Hysteresis loops of films deposited at different substrate temperature and annealed at 600 °C.

3.2. Effect of Ar gas Pressure

We observed considerable variation in the microstructure and magnetic properties of the Nd-Fe-B films while varying Ar gas pressure from 1 to 17 Pa under fixed sputter power and deposition temperature (100 W and 550 °C, respectively). XRD patterns of the films deposited under 1, 10 and 17 Pa are shown in figure 4. For all the films, c-axis texturing could be obtained; however the intensities and sharpness of the peaks are weak for low Ar pressure and phases of Nd-rich and Nd_xO_y get strengthened at high Ar pressure. Surface morphologies of these films, observed by SEM (images are not shown), indicated the grain size increases with increasing Ar pressure due to the fact that composition in the films vary accordingly. EDX analysis of the films deposited at different Ar pressures with same target composition indicates remarkable changes in the film composition (see Table 1). Table 1 shows the increase of ratio between Nd and Fe (Nd/Fe=0.21-0.31) for increasing Ar pressure. Liu *et al* [8] and Parhofer *et al* [16] showed the same trend while increases the ambient pressure. The Ar pressure during sputter greatly influences the kinetic energy of the elements on the substrate surface and hence change of composition and grain size of the films. If the gas pressure is very high, it may also lower the kinetic energy of the species in the plasma due to their intensive collisions with gas atoms, which resulted in the further drop of Nd/Fe (0.26) value for 17 Pa.

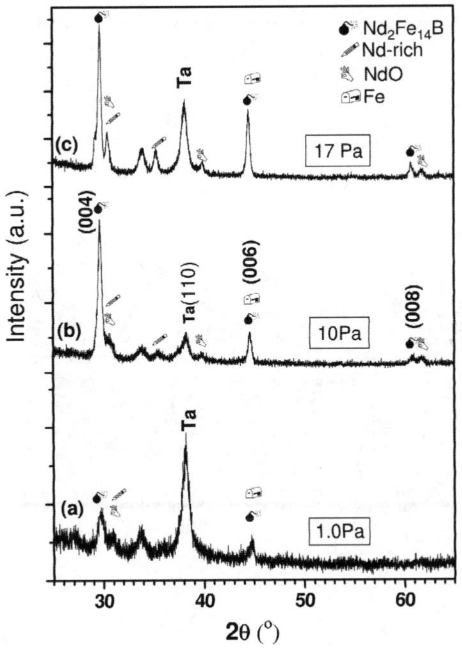

Figure 4. XRD patterns of the Nd-Fe-B films prepared under various Ar pressure.

Table 1. Ratios of Nd and Fe in the film deposited under various pressures are listed.

Ar (Pa)	Nd/Fe	Ar (Pa)	Nd/Fe
1.0	0.21	10.0	0.28
2.5	0.24	12.5	0.31
5.0	0.25	17.0	0.26
7.5	0.26		

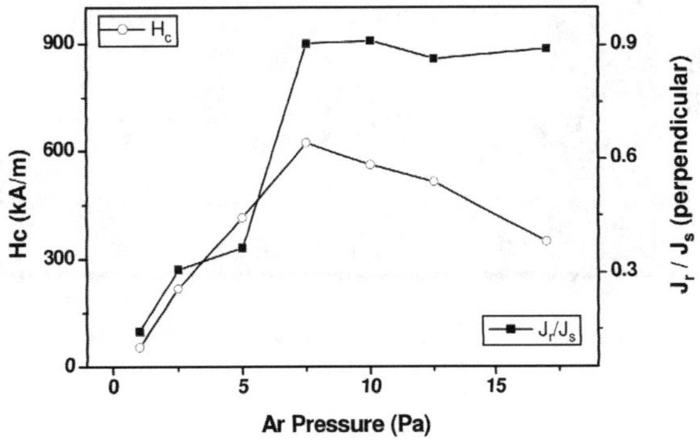

Figure 5. Ar pressure dependence on the coercivity and reduce romance of film measured along perpendicular to the film plane.

The effect of ambient gas pressure on magnetic properties of the Nd-Fe-B films is illustrated in figure 5. The reduced remanence i.e., ratio of remanence and saturation magnetization (J_r/J_s) is very low for the film prepared at very low Ar pressure. For the pressure up to 7 Pa, the reduced remanence increases and reaches close to unity and then remains almost constant for further increase of Ar pressure. It seems that the texturing of the films is better for the conditions of Ar pressure from 7 to 17 Pa. This is consistent with the result demonstrated by the x-ray diffraction data in figure 4. It is of note that Liu et al [8] observed decrease of remanence polarization with increasing Ar pressure that is different for our case. We presume that the properties of the films under study largely depend on the processing parameters such as target composition, sputter rate, etc. The increase in coercivity of the films grown at Ar pressures between 1 and 7 Pa is due to the Nd-rich phase formation. Further decrease of coercivity for increasing pressure beyond 7 Pa may perhaps be due to: (1) better alignment of the $Nd_2Fe_{14}B$ grains which makes easier magnetization reversal of whole layer once a reversed domain is formed, (2) larger grain size for which the domain wall movement occur considerably at low applied field, and (3) formation of the oxidized Nd phase.

3.3. Effect of sputter power

The effect of RF power on magnetic properties of Nd-Fe-B films was studied since the deposition rate is effectively changing according to sputter power that plays an important role in determining the magnetic behavior [9, 17, 18]. Nd-Fe-B films of about 500 nm thickness were deposited at different sputter power (50, 75, 100 and 150 W) while the substrate temperature and Ar pressure were fixed, respectively, at 550 °C and 7 Pa. The deposition rates of the films varied between 14 and 22 nm/min for different RF powers used. The surface morphology of films grown at various RF powers is depicted in figure 6. It can be clearly noticed that drastic variation in grain size as well as in shape for each RF power. Closely packed spherical shapes with average grain size of about 150 nm and island like spherical shape with average grain size of ~400 nm were observed for 50 and 75 W, respectively. For 100 W, platelet shape of 500 nm size was revealed. Rectangular shape with different size in different orientation could be obtained for the film prepared at 150 W. The increase of deposition rate will raise the Nd/Fe ratio which was already proved by Chen et al [18]. The rising values of Nd/Fe ratio were verified by EDX and they are 0.21, 0.24, 0.26 and 0.29 for 50, 75, 100 and 150 W, respectively. Nd_xO_y reflections observed by XRD are higher for the film deposited at 150 W and hence we expect that the presence of bimodal gains in it. The difference in gains is not only due to the variation in Nd content but also due to the difference in crystalline growth by bombardment of different energy particles emitted from the target [9].

Figure 6. SEM images of films deposited at various sputter power.

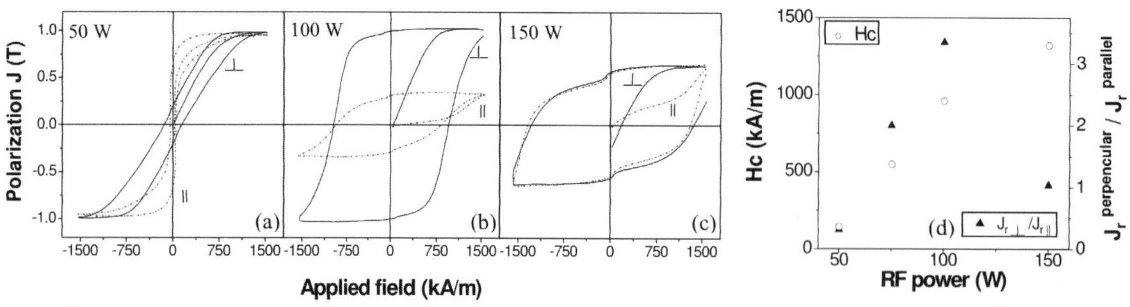

Figure 7. Hysteresis loops of the films deposited at different sputter power (a, b and c) and dependence of magnetic properties on power (d).

Figures 7 a-c show the typical hysteresis loops measured for the films grown at various power. Film sputtered at low power exhibits less hard magnetic behavior as it has less Nd content and poor anisotropic. For the power 100 W, the remanence polarization, $J_{r\perp}$, is much higher than $J_{r\parallel}$. It implies the existence of texture along c-axis which is perpendicular to film's plane. But for 150 W, the $J_{r\perp}$ and $J_{r\parallel}$ are almost same since the perpendicular anisotropic is deteriorated. The low saturation magnetization is due to the formation of non-magnetic phase as observed by XRD. Sputter power dependence on coercivity, H_c, and ratio between $J_{r\perp}$ and $J_{r\parallel}$ are shown in figure 7d. The coercivity is rising almost linearly with increasing the RF power; this is because of coexistence of the hard magnetic $Nd_2Fe_{14}B$ phase together with Nd-rich intergranular phase in the film. It is quite obvious that coercivity is sensitive to the composition of the film and high Nd concentration in the film results to the high coercivity [18]. Increase in the ratio of $J_{r\perp}$ and $J_{r\parallel}$ up to 100 W shows the increase of perpendicular anisotropic behavior and it decreases for further increase of power indicates the isotropic phase formation.

3.4. Effect of film thickness

Figure 8 plots coercivity, H_c, and reduced remanence (in perpendicular direction), J_r/J_s, with respect to various film thicknesses of samples grown at 550 °C, 7.0 Pa and 100 W. The trend of J_r/J_s shows that the texturing is better from 50 nm and above. The maximum coercivity (~1130 kA/m) was observed for a 100 nm thin film. It has been reported that the coercivity is not only influenced by the texture, but on grain size as well [5]. There may be an optimum grain size of $Nd_2Fe_{14}B$ with high coercivity for the film prepared by RF sputtering. For very thin film, the higher possibility of relative volume percentage of oxidized or defect layer hinters the nucleation field of the hard magnetic phase [10]. The H_c decrease with increasing the thickness from 100 nm, which is due to the increased grain size that was observed by SEM (image is not shown here). If the grain size is very large, the exchange coupling between soft and hard magnetic phase is tough and the domain wall motion will be easy and thus the films will have comparatively low coercivity. Figure 10 shows typical hysteresis loops for the film of 100 nm thickness and measured in the direction of perpendicular and parallel to film plane. For this condition, the Nd-Fe-B film with the values of H_c = 1130 kA/m, J_r =1.2 T and $BH_{(max)}$ = 236 kJ/m^3 in perpendicular direction were obtained.

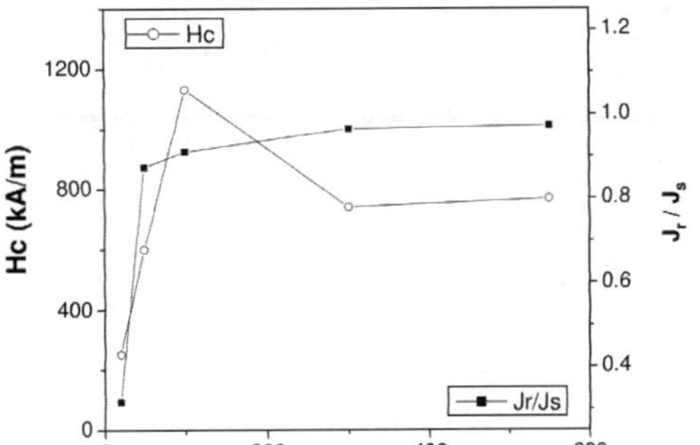

Figure 8. Influence of Nd-Fe-B film thickness on coercivity and reduced remanence.

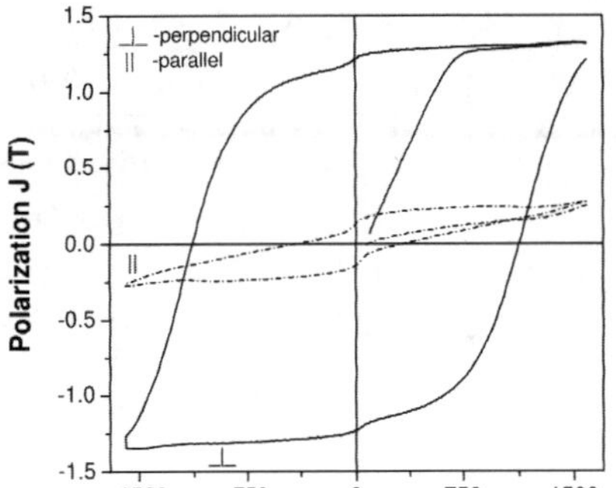

Figure 9. Hysteresis loops for a 100 nm thin Nd-Fe-B film grown at 550 °C, 7.0 Pa and 100 W.

4. Conclusion

We have prepared Nd-Fe-B thin films to study the correlations between deposition parameters of RF sputtering system and their microstructure and magnetic properties. The effect of deposition temperature on the properties of films suggested that amorphous Nd-Fe-B films grew up to 400 °C, and film grown at 500 °C had better phase formation with good texture. The Nd/Fe ratio in the film was adjustable by varying the Ar pressure and RF power. Microstructure analysis revealed that the gain size and shape are greatly influenced by RF power. Magnetic results indicated that the Nd-rich films led to higher coercivity. Excellent magnetic properties with a romance of 1.1 T and a coercivity of 960 kA/m were achieved when the film grown at a moderate RF power of 100 W. Films exhibited deterioration in their anisotropic behavior for RF power >100 W. The maximum energy product as high as 236 kJ/m^3 was obtained for a 100 nm film deposited at 550 °C under 7.0 Pa, Ar pressure and with a RF power of 100 W.

Acknowledgement
This investigation was supported by "Academic Frontier" Project for Private Universities: matching fund subsidy from MEXT (Ministry of Education, Culture, Sports, Science and Technology), 2006-2010.

References:
[1] Chin T -S 2000 *J. Magn. Magn. Meter.* **209** 75
[2] Mapps D J, Chandrasekhar R, O'Grady K, Camdridge J, Petford A and Doole R 1997 *IEEE Trans. Magn.* **33** 3007
[3] Belmans R and Hameyer K 1998 *Proceedings of the Sixth International conference on New Actuators* (Bremen, Germany) p 537
[4] Gutfleisch O 2000 *J. Phys. D: Appl. Phys.* **33** R157
[5] Parhofer S, Kuhrt C, Wecker J, Gieres G and Schultz L 1998 *J. Appl. Phys.* **83** 2735
[6] Kwon A R, Fähler S, Neu V and Schultz L 2006 *J. Magn. Magn. Meter.* **302** 252
[7] Tang S L, Gibbs M R J , Davies H A, Mateen N E, Nie B and Du Y W 2008 *J. Appl. Phys.* **103** 07E113
[8] Liu W F, Suzuki S and Machida K 2007 *J. Magn. Magn. Mater.* **308** 126
[9] Piramanayagam S N, Matsumoto M and Morisako A 1999 *J. Appl. Phys.* **85** 5898
[10] Huang X J, Xu S Y, Ong C K, Yang Z, Si L and Li Y 2002 *J. Appl. Phys.***91** 4666
[11] Chen S L, Zheng J G, Liu W and Zhang Z D 2007 *J. Phys. D: Appl. Phys.* **40** 1816
[12] Hannemann U, Fähler S, Oswald S, Holzpfel B and Schultz L 2002 *J.Magn. Magn. Mater.* **242** 1294
[13] Minowa T, Shimao M, Honshima M 1991 J. Magn. Magn. Mater. **97** 107
[14] Yue M, Zhang J X, Liu W Q, Wang G P 2004 J. Magn. Magn. Mater. **271** 364
[15] Serrona L K E B, Sugimura A, Adachi N, Okuda T, Ohsato H, Sakamoto I, Nakanishi A, Motokawa M, Ping D H and Hono K 2003 *Appl. Phys. Lett.* **82** 1751
[16] Parhofer S, Gieres G, Wecker J and Schultz L 1996 *J. Magn. Magn. Mater.* **163** 32
[17] Cadieu F J, Cheung T D, Wickramasekara L and Kamprath N 1986 *IEEE Trans. Magn.* **22** 752
[18] Chen S L, Liu W and Zhang Z D 2006 *J. Magn. Magn. Mater.* **302** 306

Journal of Physics: Conference Series **191** (2009) 012025 doi:10.1088/1742-6596/191/1/012025

Magnetic Properties of Nd-Fe-B-Zr Bulk Nanocomposite Magnets Prepared by Spark Plasma Sintering

K Tanaka[1], T Fukuzaki[2], K Nishio[1] and R Tamura[1,2]

[1]Department of Materials Science and Technology, Tokyo University of Science
[2]Polyscale Technology Research Center, Tokyo University of Science

E-mail: tamura@rs.noda.tus.ac.jp

Abstract. We have investigated the magnetic property and the microstructure of $Nd_8Fe_{85}B_5Zr_2$ bulk nanocomposite magnets consolidated by Spark Plasma Sintering (SPS) method with various sintering temperatures. The volume fractions of a soft magnetic phase (α-Fe) and a hard magnetic phase ($Nd_2Fe_{14}B$) are found to vary sensitively with the quenching speed, i.e., the surface velocity of the Cu wheel, as well as the sintering temperature. The optimal magnetic property of $(BH)_{max}$=100 kJ/m^3 has been obtained when sintered at 650 ^{o}C with the surface velocity of 25 m/s, and the high performance is attributed to a combined effect of grain refinement and reduction of non magnetic regions such as grain boundaries.

1. INTRODUCTION

Nanocomposite magnets are composed of hard and soft magnetic phases which are coupled by exchange interaction. They are expected to have a high potential as permanent magnets because of their high magnetic flux density arising from the soft magnetic phase such as α-Fe [1][2]. In particular, Nd-Fe-B nanocomposite magnets have a great advantage over the conventional sintered magnets prepared at compositions near stoichiometric $Nd_2Fe_{14}B$ with respect to the production cost because of their low rare earth content. One successful application of the nanocomposite magnets is bond magnets, which are synthesized by heat treatment of amorphous powders prepared by the melt spinning technique. However, the volume density of the bond magnets is not high enough, about 80 % of the full density, because of the unavoidable usage of binders for the formation of bond magnets. For this reason, it is preferable to synthesize full density bulk nanocomposite magnets in order to substantially improve the magnetic performance such as the magnetic flux density.

Spark Plasma Sintering (SPS) method is a unique technique which allows us to obtain full-density bulk nanocomposite magnets in a short time. In SPS, powder material is consolidated into bulk by simultaneous application of electrical current and pressure. Its advantages over other sintering techniques exist in its high sintering speed and low sintering temperature, which inhibits an excessive grain growth during sintering, making it possible to obtain a bulk sample in an amorphous or nano grains state.

Recently, various bulk nanocomposite magnets have been synthesized by using SPS, where the hard phase is $Nd_2Fe_{14}B$ and the soft phase is either α-Fe or Fe_3B [3][4]. In the case of α-Fe/$Nd_2Fe_{14}B$ nanocomposite magnets prepared by annealing melt-spun ribbons, the addition of Zr has been found to be very effective in improving the glass forming ability and, as a result, the coercivity of nanocomposite magnets has increased greatly due to the reduction of the grain size [5]. More recently, the influence of SPS sintering conditions such as the sintering temperature and the heating speed has been systematically investigated [6]. Following the previous works, the role of sintering conditions such as the sintering temperature, the quenching speed, i.e., the surface velocity of Cu wheel (v_s), in the magnetic properties and the microstructure were investigated in detail for Nd-Fe-B-Zr nanocomposite magnets prepared by the SPS method in this study.

2. EXPERIMENTAL

Alloy ingots with nominal composition $Nd_8Fe_{85}B_5Zr_2$ were prepared by melting pure elements in an arc furnace and the ingots were then melt spun onto a copper wheel rotating with the surface velocities (v_s) of 12-37 m/s. The melt spun ribbons were then crushed into powder, packed into a carbon die, and consolidated by SPS in vacuum under the applied pressure of 100 MPa . For the sintering conditions, the sintering temperature was varied between 500 and 750 °C and the heating speed was set to be 250 °C/min. After reaching the sintering temperature, the specimens were furnace-cooled with no holding time. Characterization of the phases was performed by X-ray diffraction (XRD) using Cu-Kα and the magnetic properties of the specimens were measured by using a vibrating sample magnetometer (VSM) in the magnetic field up to 2 T. The shape of the bulk samples is a disc of 10 mmφ×2 mm. The microstructure was investigated by using a field emission scanning electron microscope (FE-SEM).

3. RESULTS AND DISCUSSION

Figure 1 shows XRD patterns of $Nd_8Fe_{85}B_5Zr_2$ melt spun ribbons with different surface velocities (v_s = 12-37 m/s) together with a calculated diffraction pattern of $Nd_2Fe_{14}B$ on the bottom. For v_s = 12-25 m/s, the peaks marked by solid circles are attributed to those from α-Fe and the positions of the remaining peaks are in agreement with those of $Nd_2Fe_{14}B$. The volume fraction of α-Fe is found to increase with decreasing the surface velocity v_s. On the other hand, the melt spun ribbon prepared with v_s = 37 m/s is found to be in a nanocrystalline state.

Figure 2 presents FE-SEM back scattering images of $Nd_8Fe_{85}B_5Zr_2$ melt spun ribbons with v_s =12-37 m/s. In figure 2(a), the dark regions represent α-Fe and its grain size is estimated to be about 50-100 nm from the contrast observed in the figure. It is seen that the grain size becomes finer with increasing the surface velocity. For v_s = 18 m/s, white regions are attributed to $Nd_2Fe_{14}B$, which accounts for the high coercivity (457 kA/m) of the sample. For v_s = 25 m/s, the grain size is very fine and is not well distinguished from the image, i.e., dark and light regions are not clearly resolved on several 10 nm scale such as seen in figure 2 (a). Since both the α-Fe and the $Nd_2Fe_{14}B$ phases are recognized in the X-ray diffraction pattern (figure 1 (a)), it suggests that that the sample with v_s = 25 m/s consists of very fine grains of the soft and the hard phases.

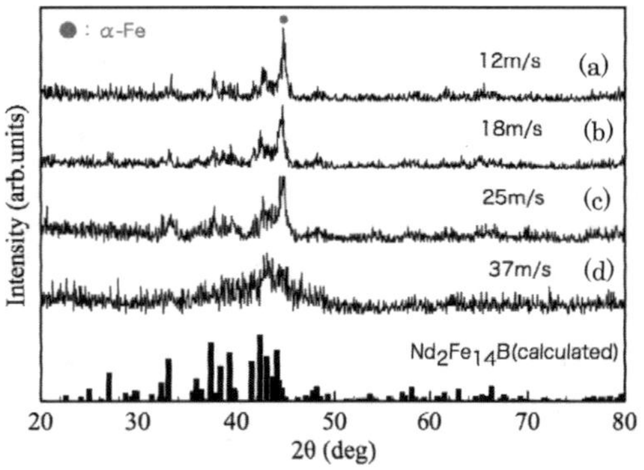

Figure1. XRD patterns of $Nd_8Fe_{85}B_5Zr_2$ melt spun ribbons prepared with the surface velocities of (a) 12m/s, (b)18m/s, (c)25m/s and (d)37m/s. On the bottom is shown the calculated diffraction pattern using the structural model of $Nd_2Fe_{14}B$.

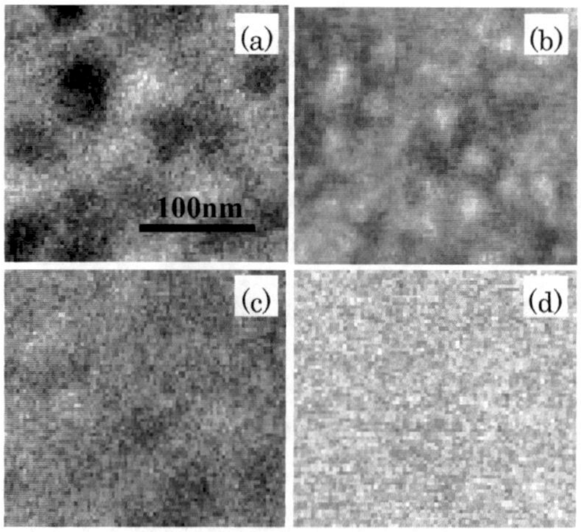

Figure2. FE-SEM back scattering images of $Nd_8Fe_{85}B_5Zr_2$ melt spun ribbons prepared with the surface velocities of (a) 12m/s, (b)18m/s, (c)25m/s, (d)37m/s.

Figure 3 shows XRD patterns of $Nd_8Fe_{85}B_5Zr_2$ bulk nanocomposite magnets prepared with v_s = 12-37 m/s together with a calculated pattern of $Nd_2Fe_{14}B$ on the bottom. As seen from the figures 3 (a)-(c), peaks from the α-Fe phase are observed in the samples consolidated at temperatures above 550 °C and the remaining broad peaks are in agreement with those of the $Nd_2Fe_{14}B$ phase. Note that no other phase is recognized from the patterns, suggesting that the samples are mostly composed of α-Fe and $Nd_2Fe_{14}B$. In addition, the relative peak intensity of the $Nd_2Fe_{14}B$ phase decreases with increasing sintering temperature, indicating that the volume fraction of α-Fe increases with increasing sintering temperatures. From the figure 3 it is also seen that the relative intensity of the α-Fe peak becomes larger for lower v_s, implying that the volume fraction of the α-Fe becomes higher for a lower quenching speed. On the other hand, for the highest quenching speed (v_s=37 m/s), the samples consolidated at 500 and 550 °C are composed mostly of the Fe_3B phase as seen from the figure 3(d), which is in contrast with the case of the sample consolidated at 600 °C where the α-Fe and the $Nd_2Fe_{14}B$ phases coexist. When consolidated above 650 °C, the peak intensity of the $Nd_2Fe_{14}B$ phase decreases relatively to those of α-Fe, showing that the volume fraction of α-Fe phase increases when consolidated at higher temperatures above 650 °C. The above results indicates that lower sintering temperatures are favorable for obtaining higher volume fraction of $Nd_2Fe_{14}B$.

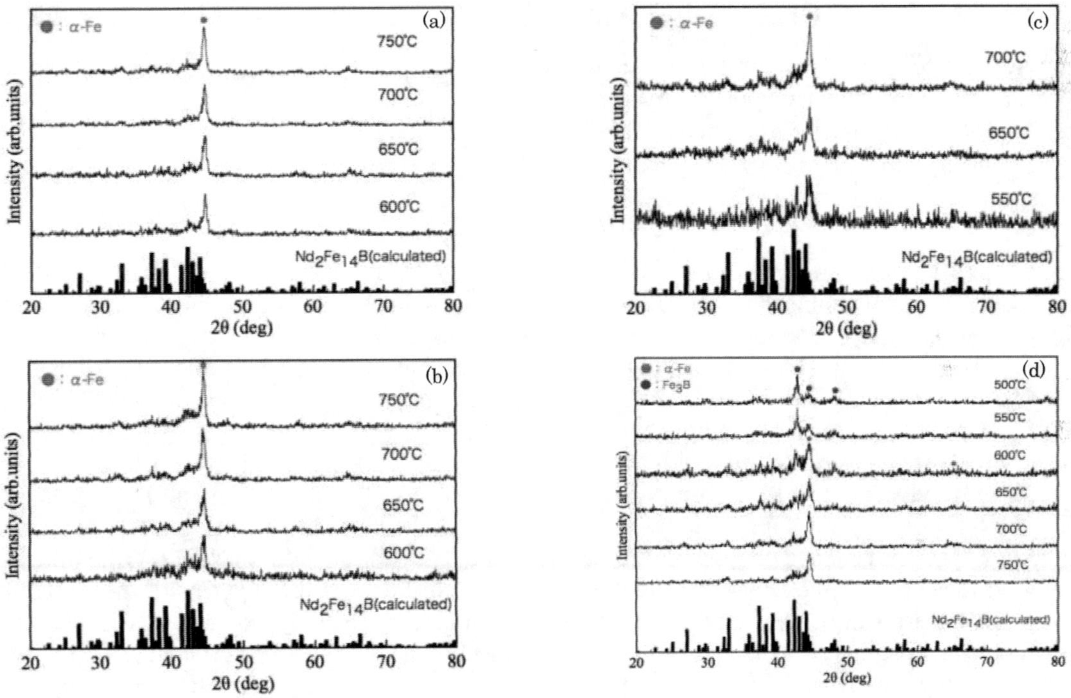

Figure3. XRD patterns of $Nd_8Fe_{85}B_5Zr_2$ bulk nanocomposite magnets prepared with the surface velocities of (a) 12m/s, (b)18m/s, (c)25m/s and (d)37m/s, and sintered at 500-750℃. On the bottom is shown the calculated diffraction pattern using the structural model of $Nd_2Fe_{14}B$.

Figure 4 presents FE-SEM back scattering images of $Nd_8Fe_{85}B_5Zr_2$ bulk magnet prepared with v_s =12-37 m/s sintered at 650 °C. The grain size of the magnet prepared with v_s =12 m/s is considerably larger than those of the magnets with the higher surface velocities v_s, and it becomes finer with increasing v_s. Figure 5 shows FE-SEM back scattering images of $Nd_8Fe_{85}B_5Zr_2$ sintered at 650 and 700 °C. The grain size of the magnet sintered at 650 °C is about 50 nm whereas that of the magnet sintered at 700 °C is about 100 nm, showing that the grain size varies sensitively with the sintering temperature. A lower sintering temperature is effective in the suppression of the grain growth during sintering.

From these observations we can conclude that the quenching speed, i.e., the surface velocity, and the sintering temperature are two important factors in determining the volume fraction of α-Fe and $Nd_2Fe_{14}B$ and the grain size of the nanocomposites.

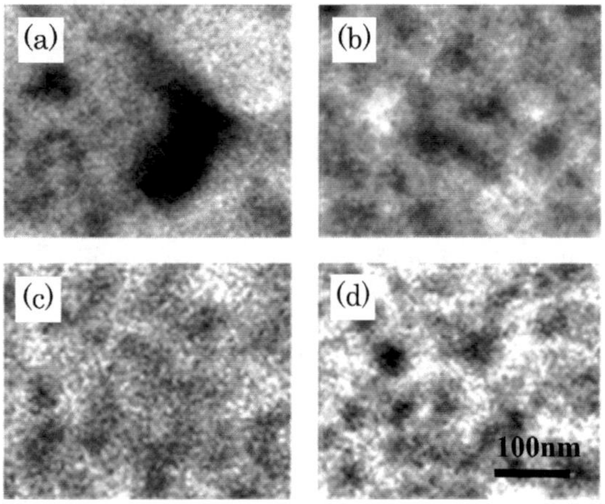

Figure4. FEM-SEM back scattering images of $Nd_8Fe_{85}B_5Zr_2$ bulk nanocomposite magnets sintered 650 ℃ prepared with the surface velocities of (a) 12m/s, (b)18m/s, (c)25m/s and (d)37m/s.

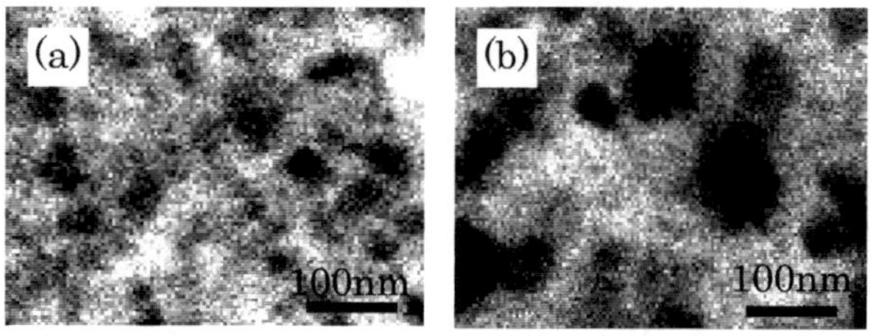

Figure5. FE-SEM back scattering images of $Nd_8Fe_{85}B_5Zr_2$ bulk nanocomposite magnets prepared with the surface velocity of 25m/s and sintered at (a)650 °C and (b)700 °C

Figure 6 presents sintering temperature dependence of the maximum energy product $(BH)_{max}$ of the $Nd_8Fe_{85}B_5Zr_2$ bulk nanocomposite magnets prepared with v_s =12-37 m/s. The sample sintered at 650 °C exhibits highest energy product $(BH)_{max}$ among those prepared with the other temperatures for all the surface velocities (v_s =12-37 m/s). The high $(BH)_{max}$ for the 650 °C sintered samples is attributed to strong exchange coupling effect between the soft and hard grains since the grain growth is effectively suppressed for the 650 °C–sintered samples. Figure 7 shows M-H curves of the bulk magnets sintered at 650 °C with different surface velocities. The highest saturation magnetization of 1.38 T and the highest energy product of 100 kJ/m^3 are obtained for v_s =25 m/s. The reason for the best magnetic performance for v_s =25 m/s is considered to be twofold. One is the high Mr/Ms ratio of 0.75, which is attributed to effective exchange coupling effect between the soft and the hard phases, and the other is high coercivity, which indicates that the volume fraction of $Nd_2Fe_{14}B$ is the highest for the sintering condition. The latter result means that the volume fraction of $Nd_2Fe_{14}B$ can be controlled by tuning the surface velocities. The present work has shown that the quenching speed i.e. the surface velocity of the Cu wheel, is a very important factor to determine the volume fraction of nonmagnetic regions such as grain boundaries, the volume fraction of $Nd_2Fe_{14}B$ and the grain size.

It is noted that concerning the saturation magnetization, one might expect much higher saturation magnetization exceeding 1.6 T for nanocomposite magnets entirely composed of α-Fe and $Nd_2Fe_{14}B$ since the saturation magnetization is determined by the volume fraction and the saturation magnetization of the constituent phases,. However, the observed saturation magnetizations are less than 1.6 T for all the samples, suggesting that the samples sintered at 650 °C with v_s =12-37 m/s contain either nonmagnetic or weak magnetic regions within the specimens. Such a phase has not been recognized in the present work, however, grain boundaries might be responsible for the loss of the magnetization. The present results suggest that the volume fraction of nonmagnetic regions is the lowest for the bulk magnet prepared with v_s = 25 m/s and sintered at 650 °C.

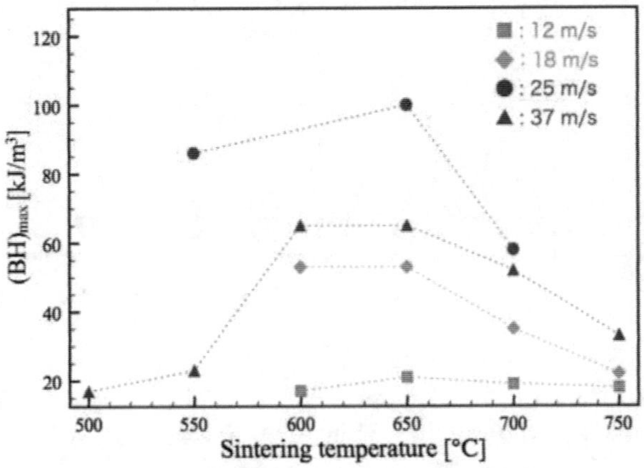

Figure6. Sintering temperature dependence of $(BH)_{max}$ for the $Nd_8Fe_{85}B_5Zr_2$ bulk nanocomposite magnets prepared at v_s =12-37m/s

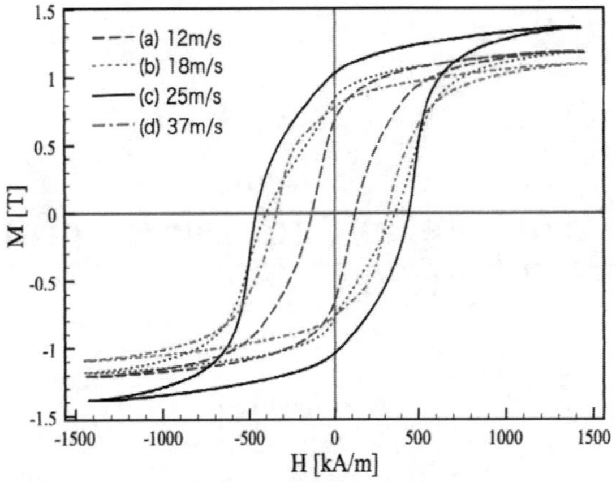

Figure7. Magnetization curves of the $Nd_8Fe_{85}B_5Zr_2$ bulk nanocomposite magnets sintered at 650℃ with the various surface velocities of (a) 12m/s, (b)18m/s, (c)25m/s and (d)37m/s.

4. CONCLUSIONS

We have investigated the magnetic properties and the microstructure of $Nd_8Fe_{85}B_5Zr_2$ bulk nanocomposite magnets consolidated by spark plasma sintering (SPS) method at various temperatures as well as with various quenching speeds. It has been demonstrated that the grain size and the volume fraction of $Nd_2Fe_{14}B$ and weak or non magnetic regions such as grain boundaries can be controlled by the sintering temperature and the quenching speed, respectively. As a result of the optimization by the two sintering conditions, the highest energy product of $(BH)_{max}$=100 kJ/m^3 has been obtained at the surface velocity of 25 m/s and the sintering temperature of 650 °C for the $Nd_8Fe_{85}B_5Zr_2$ nanocomposite magnet.

5. ACKNOWLEDGMENTS

This work was supported by "Academic Frontier" Project for Private Universities: Matching Fund Subsidy from MEXT (Ministry of Education, Culture, Sports, Science and Technology), 2006-2010.

REFERENCES

[1] Kneller E F and Hawig R 1991 *IEEE Trans. Magn.* **27** 3588
[2] Skomski R and Coey J M D 1993 *phys. Rev.*B **48** 15812
[3] Yue M, Zhang J and Tian M 2006 *J.Appl.Phys.* **99** 08B502-1
[4] Saito T, Takeuchi T, and Kageyama H 2004 *IEEE Trans Magn.***40** 2880
[5] Wu Y Q, Ping D H, Xiong X Y and Hono K 2002 *J.Appl.Phys* **91** 8174
[6] K Tanaka, T Fukuzaki et al 2008 Journal of Physics Conference Series **106** 012014

Structural and magnetic characteristics of SrM thin films on Au dot arrays

A Kaewrawang[1], A Ghasemi[1], X Liu[1,]* and A Morisako[1]
[1]Spin Device Technology Center, Faculty of Engineering, Shinshu University, 4-17-1 Wakasato, Nagano City, Nagano 380-8553, Japan.

*E-mail: liu@cs.shinshu-u.ac.jp

Abstract. The self-assembled SrM dot array on Au islands has been prepared at various substrate temperatures for underlayer by using DC magnetron sputtering system. *C*-axis oriented SrM perpendicular to the films can be observed only at substrate temperature for underlayer of room temperature. The intensity of (111) diffraction line for Au and that of (00*l*) diffraction line for strontium ferrite decreases with increasing substrate temperature for underlayer. The maximum of coercivity and remanent squareness ratio in perpendicular direction are 3.4 kOe and 0.6, respectively, at substrate temperature for underlayer of room temperature.

1. Introduction

The areal density of hard disk drives continues to increase. However, the trend will be limited by thermal instability or superparamagnetism of the grains in medium. Patterned media [1,2], in which data are stored in isolated single-domain islands, have been one of the widely discussed possibilities for extending data densities with its capability to obtain higher storage densities well in excess of 1 Tbit/in^2.

There are several methods to fabricate patterned films e.g. lithography [3], irradiation through a mask [4], milling with focused ion beam [5] and imprint technology [6]. Self-assembly prepared by sputtering deposited method is used because it has the potential for making very fine-scale structures over large areas in a rapid process and no chemical used in process.

The requirements of magnetic material for patterned media are high coercivity, chemical stability and perpendicular anisotropy. Hexagonal strontium ferrite (SrM) has fairly large magnetocrystalline anisotropy, excellent chemical stability, magnetic moment aligned along the *c*-axis and minimum stable grain size about 8.8 nm which is computed for a storage time constant 10 years using the media stability criterion $KV > 60k_BT$. Due to these characteristics, SrM is an attractive material for patterned media application.

2. Experiments

The SrM film was deposited on 10 nm thickness of Au underlayer using a magnetron sputtering apparatus. The Corning glass was used as substrate and thickness of SrM layer was kept at 30 nm. After evacuating the chamber to a pressure below 2×10^{-6} Torr, argon (Ar) was introduced and sputtering pressure was set at 2.0 mTorr for Au underlayer film. In order to get Au island structure, the substrate temperature for Au underlayer, T_u, was varied in the range from room temperature (r.t.) to 300 °C. A sintered ferrite disk with stiochiometric composition of M-type, i.e. Sr:Fe = 1:12, was used as the target for SrM film. The mixture gas of Ar and O_2 with gas pressure of 1.98 and 0.02 mTorr, respectively, was used as sputtering gas. The substrate temperature for SrM layer, T_s, was kept constant at 525 °C. The crystallographic properties of the films were characterized by an x-ray diffractometer (XRD) with Cu K_α radiation. The magnetic properties were measured by vibrating sample magnetometer (VSM) with maximum field strength of 24 kOe. The surface morphologies of the films were observed using scanning electron microscope (SEM) and atomic force microscope (AFM). The elemental composition of the films was analyzed by using energy dispersive x-ray spectroscope (EDS).

3. Results and Discussion

The Au film is expected to be used as the underlayer for promotion of *c*-axis orientation of hexagonal ferrite layer. The misfit ratio between (111) plane of *fcc* Au and *c*-plane of *hcp* SrM is 2.1%. Figure 1 shows the Au surface prepared on Corning glass at various substrate temperatures, T_u. The Au island structures can be formed at T_u higher than 100 °C. The island size distributions for Au are shown in Fig. 2. As the growth temperature increases, the curves shift to smaller diameters meaning that the number of small dots increases while the number of large ones decreases. The average Au dot size is about 25.8, 22.9 and 20.4 nm at T_u of 100, 200 and 300 °C, respectively.

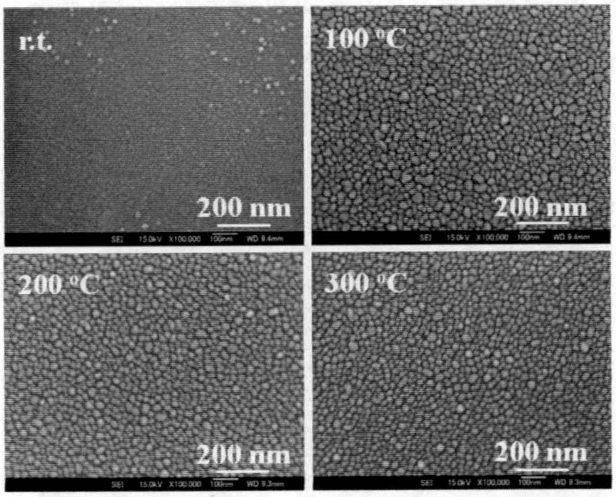

Figure 1. Plain view of Au underlayer on Corning glass at various T_u.

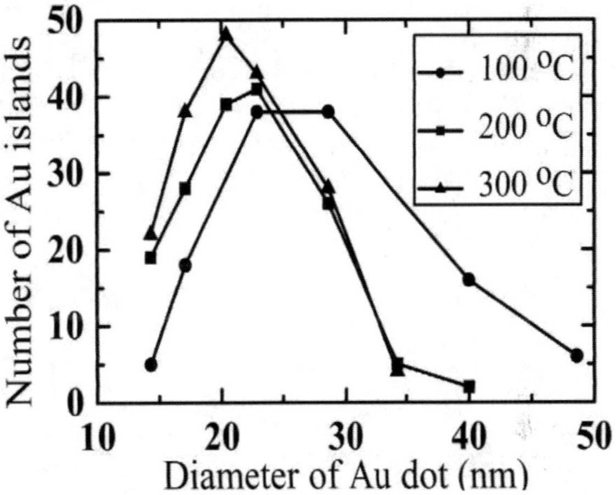

Figure 2. Distribution of Au island size in 0.13 μm^2.

Figure 3 shows the dependence of saturation magnetization, M_s, of SrM films on substrate temperature, T_s with and without Au underlayer. We have found that the SrM films deposited on Au underlayer can be crystallized at T_s above 515 °C. On the other hand, the SrM deposited directly on Corning glass can not be crystallized when the substrate temperature is as high as 550 °C. It has reported that the infrared can be absorbed by Au particles [7]. It may contribute the crystallization of SrM. From these results, an attempt to prepare the isolated SrM patterned films has been carried out by using Au underlayer. The proposed cross sectional structure in this study is schematically shown in Fig. 4.

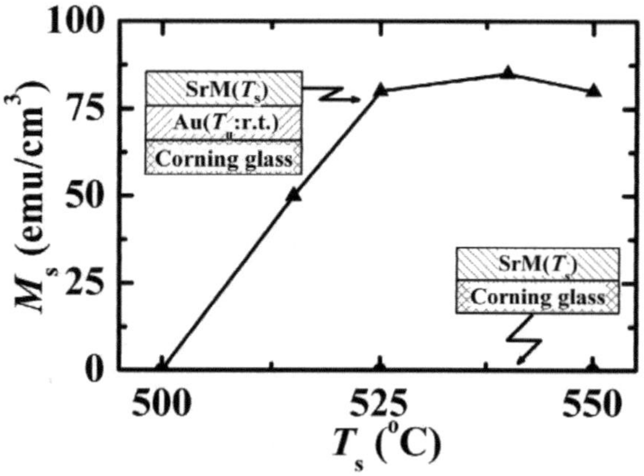

Figure 3. Dependence of M_s on substrate temperature, T_s.

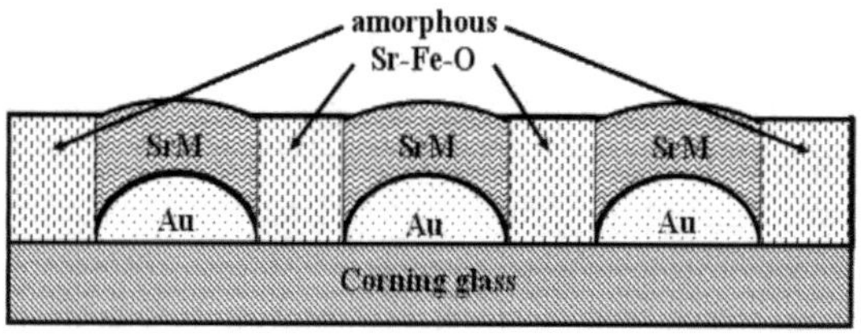

Figure 4. Schematic of cross sectional view of partially crystallized SrM.

Figure 5 shows XRD diagrams for SrM/Au films prepared at various T_u. The preferred orientation of (111) *fcc* lattice of Au underlayer can be observed for all of the films. With increasing T_u as shown in Fig. 6, the intensity of diffraction line from Au(111) peak becomes weak, the full width at half maximum (FWHM) of rocking curve, $\Delta\theta_{50}$, of Au(111) increases from 1.9° to 14.9°. On the other hand, the Au(200) peak increases with increase of T_u. Consequently, the diffraction lines from *c*-plane of hexagonal SrM phase have been observed for the films prepared at T_u of r.t. and it becomes weaker with increasing T_u. There are no obvious diffraction lines from SrM phase for T_u above 50°C. These results indicate that the (111) orientation of Au underlayer promotes *c*-axis oriented SrM.

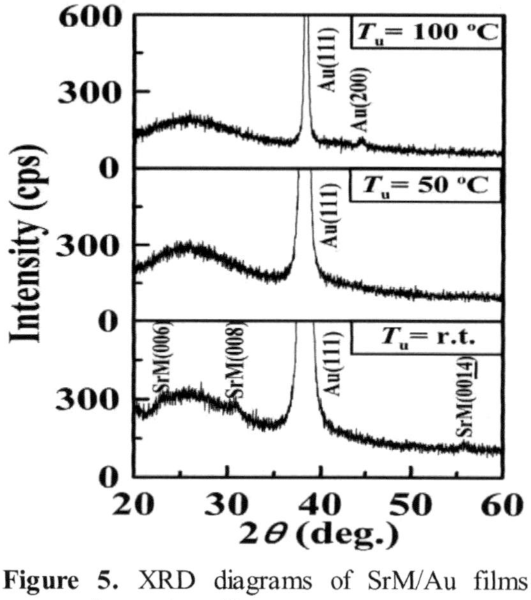

Figure 5. XRD diagrams of SrM/Au films prepared at various T_u.

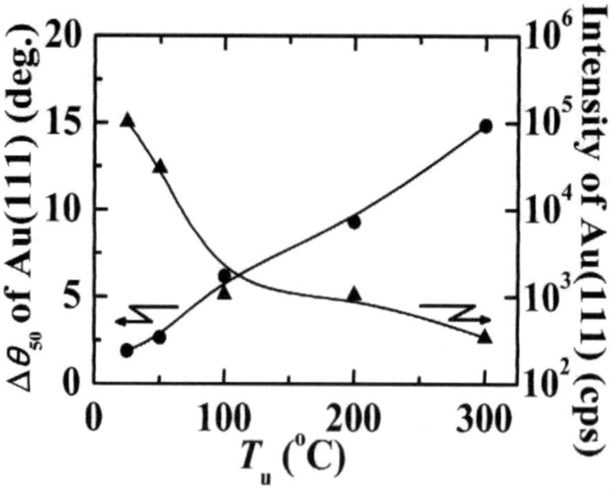

Figure 6. $\Delta\theta_{50}$, of Au(111) and Intensity of Au(111) as function of T_u.

As shown in Fig. 7, the coercivity in perpendicular direction, $H_{c\perp}$, is higher than that for in-plane direction, $H_{c//}$. It means that all of the films exhibit perpendicular anisotropy. The squareness ratio, S, is almost same trend as coercivity. $H_{c\perp}$ and $S_\perp$ decrease from 3.4 to 1.0 kOe and from 0.6 to 0.3, respectively, with increasing T_u from r.t. to 300 °C. The saturation magnetization, M_s, is almost constant at 90 emu/cm^3.

The interactions between magnetic particles of the thin films have been evaluated with the Δm^* method [8]. Following this method, the presence and strength of interactions is described as this equation: $\Delta m^*(H) = m_i(H)-0.5[m_u(H)+m_l(H)]$ where $m_i(H)$ is the initial magnetization curve, $m_u(H)$ is the upper branch and $m_l(H)$ is the lower branch of the hysteresis loop. The finite negative and positive values of Δm^* indicate the magnetostatic and exchange coupling, respectively.

In Fig. 8 the Δm^*(H) deviations of the SrM/Au thin films with different T_u are reported. It indicates negative interactions in the films. The peak of Δm^* is smaller in higher T_u films, which seems to indicate a tendency towards weaker interactions when the substrate temperature for Au underlayer, T_u, increases. From this results, it might be contributed from non-magnetic phase existed in the space between the magnetic particles in SrM thin films and hence the intergranular interactions is reduced.

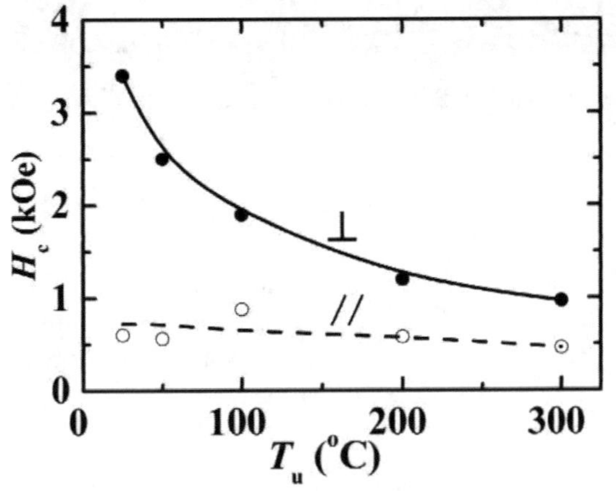

Figure 7. Variation of coercivities as a function of T_u.

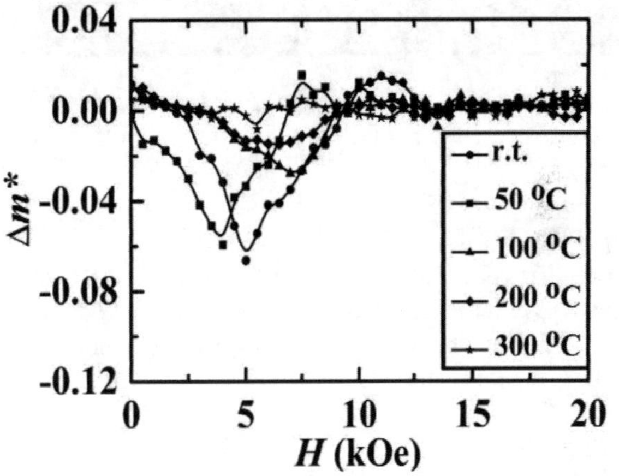

Figure 8. Δm^* deviations for SrM/Au thin films with different T_u.

Figure 9 shows SEM micrographs of Au morphology after heating the substrate at 525 °C for 25 min. It can be seen that the surface morphology of Au prepared at r.t. is changed to separate in two parts including white and black, while the others prepared at higher temperature are not changed. From EDS analysis of the Au film prepared at r.t. after that heating up to 250 °C (not be shown here), we have found that the intensity of Au in the black area is lower than white area because the amount of Au in the black section is 3 to 6 times lower than white area. From these results, we can conclude that the thickness of white part of the heating Au surface prepared at r.t. is higher than that for the black one. The grain size of SrM grain on Au underlayer gradually decreases with increasing T_u from r.t. to 300 °C as shown in Fig.10.

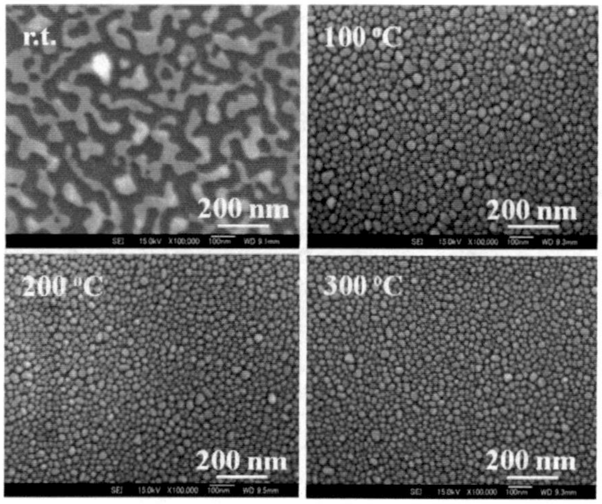

Figure 9. Au surface prepared on Corning glass at various T_u after heating the substrate at 525 °C for 25 min..

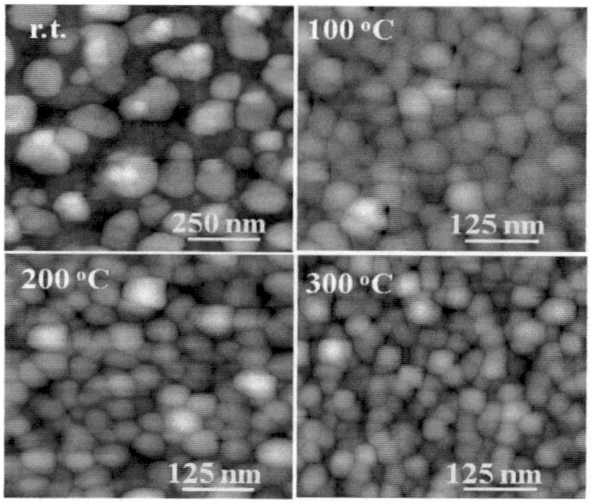

Figure 10. SrM/Au surfaces prepared at various T_u.

4. Conclusions

The Au island structure can be formed on Corning glass by heating the substrate over 100 °C. The perpendicular grown of SrM layer can be observed on Au underlayer only at T_u of room temperature. With increasing T_u, the crystallographic and magnetic properties of SrM/Au thin films are deteriorated. The magnetic interaction between SrM dot arrays prepared on Au is reduced with increasing T_u.

Acknowledgments
This work was supported in part by Grants-in-Aid for Scientific Research (B) and Grant-in-Aid for JSPS Fellows from Japan Society for Promotion of Science (JSPS).

References

[1] Chou S Y 1997 *Proc. IEEE* **85** 652
[2] White R L, New R M H and Pease R F W 1997 *IEEE Trans. Magn.* **33** 990
[3] Castano F J, Hao Y, Haratani S, Ross C A, Vogrli B, Walsh M and Smith H I 2001 *IEEE Trans. Magn.* **37** 2073
[4] Chappert C, Bernas H, Ferre J, Kottler V, Jamet J P, Chen Y, Cambril E, Devolder T, Rousseaux F, Mathet V and Launois H 1998 Science **280** 1919
[5] Rettner C T, Best M E and Terris B D 2001 *IEEE Trans. Magn.* **37** 1649
[6] Heidari B, Maximov I and Montelius L 2000 *J. Vac. Sci. Technol.* B **18** 3557
[7] Suzuki Y, Makanae H, Kudo H, Miyanaka T, Nanke T and Kobayashi T 2004 *Appl. Phys.* A **78** 335
[8] Thamm S and Hesse J 1996 *J. Magn. Magn. Mater.* **154** 254

Magnetic and Electric Properties of Mn-Zn Ferrite Fabricated by Pulsed Laser Deposition

H Etoh[1,2], J Sato[2], Y Murakami[2], A Takahashi[2] and R Nakatani[1]

[1] Division of Materials and Manufacturing Science, Graduate School of Engineering, Osaka University, 2-1 Yamadaoka, Suita, Osaka 565-0871, Japan

[2] Advanced Technology Research Laboratories Corporate Research and Development Group, Sharp Corporation, 2613-1 Ichinomoto-cho, Tenri, Nara 632-8567 Japan

E-mail: etoh-hideki@sharp.co.jp

Abstract. We investigated the basic properties of Mn-Zn ferrite thin films fabricated by pulsed laser deposition (PLD) for the purpose of controlling their saturation magnetizations (M_s) and electrical resistivities. The M_s and the electrical resistivities of fabricated samples varied with their compositions and the post-annealing conditions. It is indicated that the compositions and partial pressure of oxygen in the annealing atmosphere influence to the properties.

1. Introduction

In the field of magnetic recording, the thermal agitation limit is a serious problem for higher recording density beyond Tb/inch^2. A thermally assisted magnetic recording method is proposed to overcome the problem which use magnetic field and laser beam for recording [1]. In this method, it is important to apply the magnetic recording field and the laser beam to the same position on the recording media. Although it is effective to incorporate a magnetic core in the recording head to apply a sufficiently large recording field, the core could possibly interfere with the optical path of the laser beam. So the magnetic core for the thermally assisted magnetic recording is required to be transparent, as well as to exhibit a high M_s for enough recording field and high resistivity for reduction of eddy current loss at high-frequency magnetic field. Mn-Zn ferrite is a transparent magnetic material with a relatively high M_s and high resistivity [2], and it can be a candidate for a magnetic core that will not interrupt the laser beam. In this paper, we report basic properties of Mn-Zn ferrite thin films fabricated by PLD aiming for the magnetic core of a thermally assisted magnetic recording head.

2. Experiment

Mn-Zn ferrite thin films with the thickness of 50 nm were fabricated on thermally oxidized silicon substrates by a PLD equipment using a Nd:YAG laser with the wavelength of 266 nm and operated at 10 Hz at a vacuum pressure of 1 x 10^{-4} Pa. We used four Mn-Zn ferrite targets with the compositions of $Mn_{0.5}Zn_{0.5}Fe_2O_x$, $Mn_{0.24}Zn_{0.69}Fe_{2.1}O_x$, $Mn_{0.65}Zn_{0.65}Fe_{1.7}O_x$, and $Mn_{0.7}Zn_{0.3}Fe_2O_x$. The target and substrate were rotated during deposition. The post-annealing conditions were at a vacuum pressure of 1 x 10^{-4} Pa for 20 min and in air atmosphere for 5 minutes. Magnetic properties were measured by a vibrating sample magnetometer and crystalline structures were analyzed by X-ray Diffraction (XRD).

Journal of Physics: Conference Series **191** (2009) 012027 doi:10.1088/1742-6596/191/1/012027

Electrical resistivities were measured by the two probe method and metal compositions were measured by X-ray Fluorescence (XRF) analysis.

3. Results and discussions

3.1. X-ray diffraction measurement

Figures 1 and 2 show XRD patterns of samples deposited using a $Mn_{0.5}Zn_{0.5}Fe_2O_x$ target. Figure 1 shows the results of the samples annealed at a vacuum pressure of 1 x 10^{-4} Pa and figure 2 shows those annealed in air atmosphere. In figure 1, clear Mn-Zn ferrite peak was not observed in non-annealed sample, but in the samples annealed at and above 350°C, (3 1 1) peaks of Mn-Zn ferrite were observed. In figure 2, (3 1 1) peaks of Mn-Zn ferrite could be observed but the intensities were lower than those in figure 1. Figures 3 and 4 also show XRD patterns of samples whose target composition is $Mn_{0.24}Zn_{0.69}Fe_{2.1}O_x$. As shown in figure 4, the samples annealed in air atmosphere showed not only (3 1 1) peaks but also (1 1 1) peaks annealed at and above 400°C. The reason of appearance of (1 1 1) peaks is yet to be investigated. The samples deposited using other two targets ($Mn_{0.65}Zn_{0.65}Fe_{1.7}O_x$ and $Mn_{0.7}Zn_{0.3}Fe_2O_x$) showed similar tendency with the results shown in the figures 1 and 2.

3.2 Metal composition measurement

Figures 5 and 6 show annealing temperature dependence of metal compositions measured by XRF analysis of samples deposited using a $Mn_{0.5}Zn_{0.5}Fe_2O_x$ target. Figure 5 shows the results of the samples annealed at a vacuum pressure of 1 x 10^{-4} Pa. Figure 6 shows those annealed in air atmosphere. As shown in figure 5, the samples annealed in vacuum exhibited the decrease of Zn ratio, and increase of the Mn and Fe ratio with an increase in annealing temperature. In contrast, the samples annealed in air atmosphere (figure 6) showed little change in compositions. It is thought Zn evaporated in the samples annealed in vacuum because equilibrium vapor pressure of Zn is much higher than those of Mn or Fe at annealing temperature, but evaporation of Zn in the samples annealed in air atmosphere was suppressed.

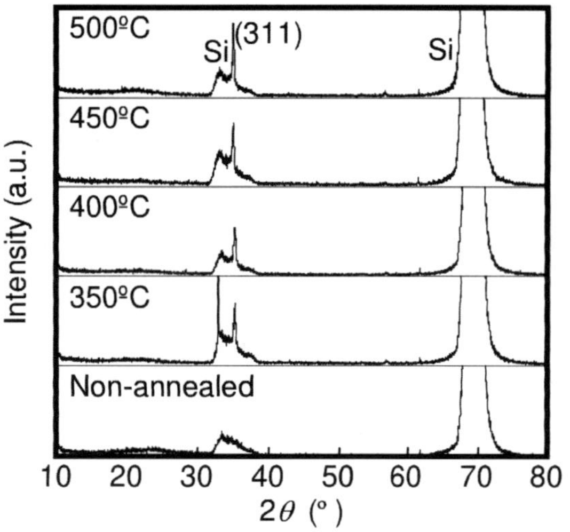

Figure 1. XRD patterns of the samples annealed at a vacuum pressure of 1 x 10^{-4} Pa (target composition of $Mn_{0.5}Zn_{0.5}Fe_2O_x$).

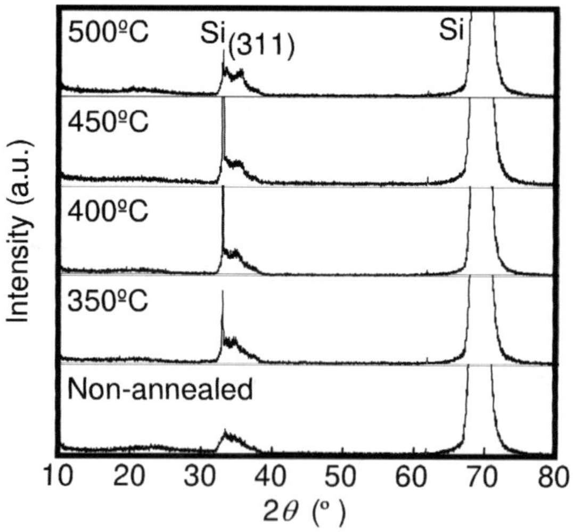

Figure 2. XRD patterns of the samples annealed in air atmosphere (target composition of $Mn_{0.5}Zn_{0.5}Fe_2O_x$).

3.3 Magnetic properties

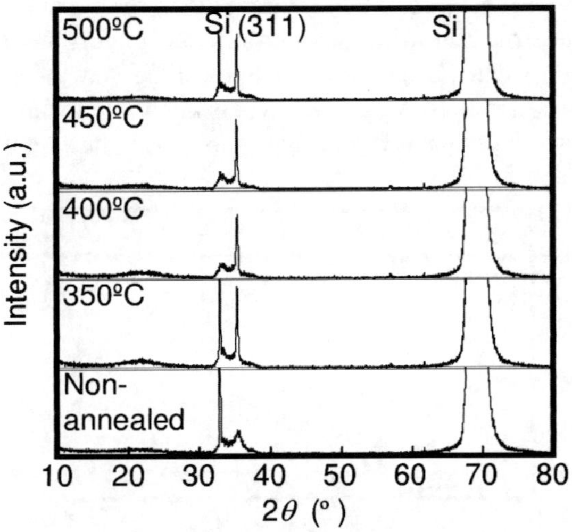

Figure 3. XRD patterns of the samples annealed at a vacuum pressure of 1 x 10^{-4} Pa (target composition of $Mn_{0.24}Zn_{0.69}Fe_{2.1}O_x$).

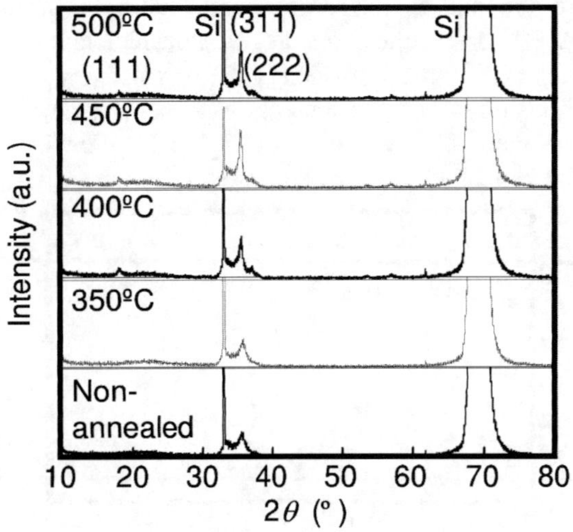

Figure 4. XRD patterns of the samples annealed in air atmosphere (target composition of $Mn_{0.24}Zn_{0.69}Fe_{2.1}O_x$).

Figure 7 (a) to (d) show the annealing temperature dependence of saturation magnetizations (M_s) for the Mn-Zn ferrite films fabricated using the four targets. In all the non-annealed samples shown in figure 7 (a) to (d), Mn-Zn ferrites showed small M_s of less than 0.03 T, and the M_s increased with annealing. All the samples annealed in vacuum showed larger M_s than those for samples annealed in air atmosphere at the same temperature. The M_s values after annealing were different depending on the compositions of the ferrites, and the maximum M_s value after annealing in vacuum and in air atmosphere were 0.58 T at the temperature of 450ºC and 0.24 T at the temperature of 400ºC, respectively. The both values were observed when the $Mn_{0.24}Zn_{0.69}Fe_{2.1}O_x$ target was used. The results that the M_s increased after annealing agree with the results of increase in intensity of (3 1 1) peak of Mn-Zn ferrite in figures 1, 2, 3, and 4. There is also thought to be a possibility that M_s values were influenced by Zn amount decrease and oxygen amount loss during the annealing for the samples annealed in vacuum.

3.4 Electric properties

Figure 8 shows the annealing temperature dependence of electrical resistivity of Mn-Zn ferrites annealed at a vacuum pressure of 1 x 10^{-4} Pa and in air atmosphere. In all the compositions, the resistivities of the samples annealed in air atmosphere increased over 700 Ωcm with an increase in annealing temperature over 400-500ºC, while those annealed in vacuum decreased with an increase in annealing temperature except for the samples of $Mn_{0.65}Zn_{0.65}Fe_{1.7}O_x$ target. The increase of resistivity of the samples annealed in air atmosphere is considered to be partly caused by formation of Fe_2O_3 which has high resistivity [3]. But intensity of XRD diffraction is very broad and weak because samples are thin, so the both peaks could not apparently distinguished. The decrease of resistivity in the samples annealed in vacuum is also thought to be attributed to the oxygen loss during annealing in vacuum.

4. Conclusions

Basic properties of four compositions of Mn-Zn ferrite thin films fabricated by pulsed laser deposition were evaluated to investigate their possible application as new magnetic cores in a thermally assisted magnetic recording. The M_s and the resistivities of the Mn-Zn ferrites were found to depend on annealing atmosphere and on metal compositions. A higher M_s of 0.58 T and a lower resistivity of 0.05 Ωcm were obtained for the samples annealed in vacuum of $Mn_{0.24}Zn_{0.69}Fe_{2.1}O_x$ target at the temperature of 450ºC, and a relatively lower M_s of 0.14 T and a higher resistivity of over 700 Ωcm

Journal of Physics: Conference Series **191** (2009) 012027 doi:10.1088/1742-6596/191/1/012027

were obtained for the samples annealed in air atmosphere of $Mn_{0.7}Zn_{0.3}Fe_2O_x$ target at the temperature of 400°C. From these results, we consider that Mn-Zn ferrite thin films with both a sufficiently high M_s (0.3 T) and a sufficiently high resistivity (10^2 Ωcm) can be obtained by controlling the metal compositions and the annealing process, especially the partial pressure of oxygen in the annealing atmosphere. This will enables the Mn-Zn ferrites as practical magnetic core materials in the thermally assisted magnetic recording system.

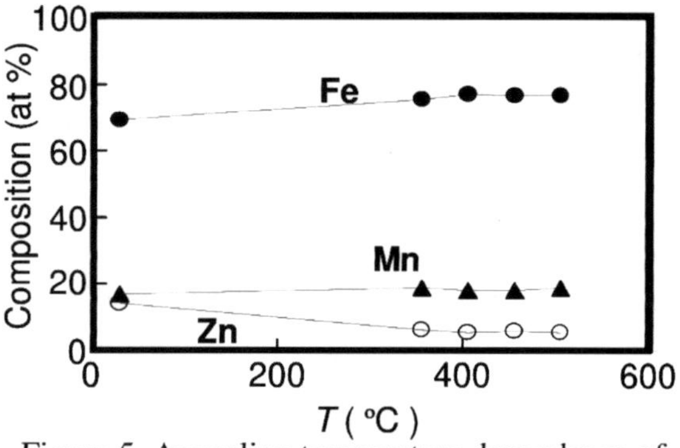

Figure 5. Annealing temperature dependence of metal compositions measured by XRF analysis of the samples (target composition of $Mn_{0.5}Zn_{0.5}Fe_2O_x$) annealed at a vacuum pressure of 1 x 10^{-4} Pa.
Closed circles, closed triangles and open circles denote Fe, Mn and Zn, respectively.

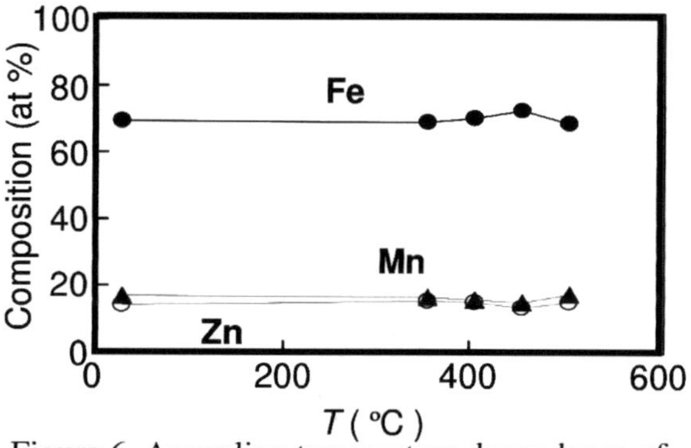

Figure 6. Annealing temperature dependence of metal compositions measured by XRF analysis of the samples (target composition of $Mn_{0.5}Zn_{0.5}Fe_2O_x$) annealed in air atomosphere. Closed circles, closed triangles and open circles denote Fe, Mn and Zn, respectively.

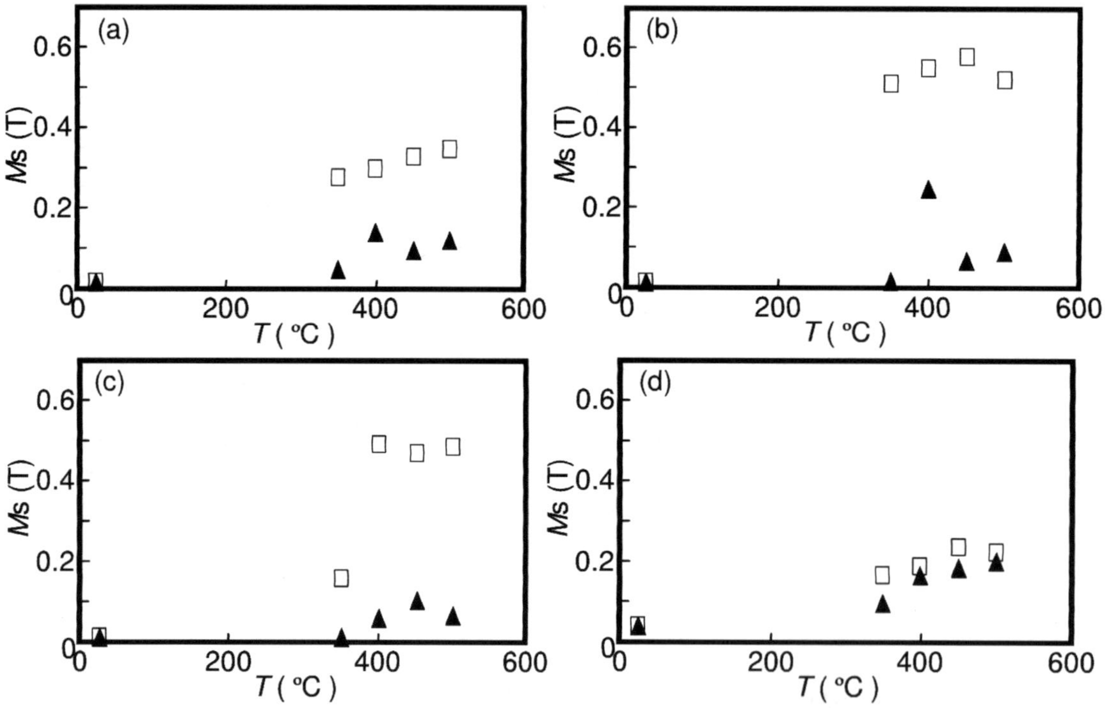

Figure 7. Annealing temperature dependence of M_s. The target compositions are (a) $Mn_{0.5}Zn_{0.5}Fe_2O_x$, (b) $Mn_{0.24}Zn_{0.69}Fe_{2.1}O_x$, (c) $Mn_{0.65}Zn_{0.65}Fe_{1.7}O_x$ and (d) $Mn_{0.7}Zn_{0.3}Fe_2O_x$. Open squares and closed triangles denote the samples annealed at a pressure of 1 x 10^{-4} Pa and in an air atmosphere, respectively.

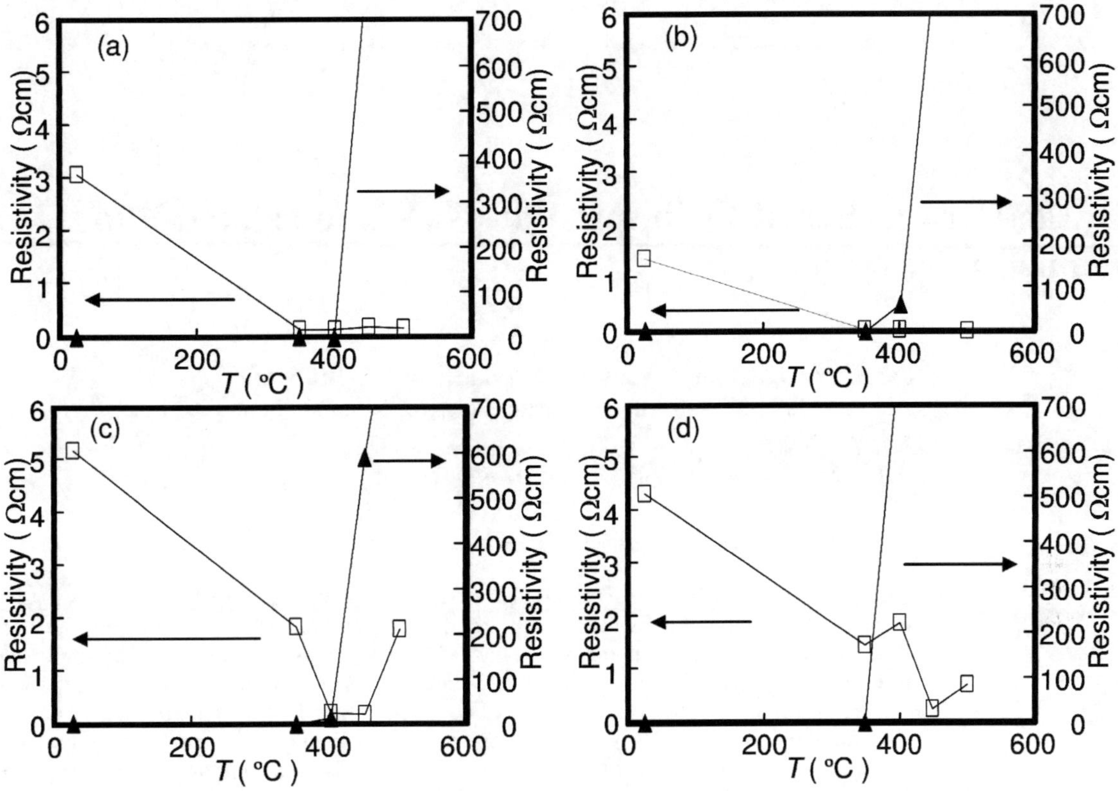

Figure 8. Annealing temperature dependence of resistivity. The target compositions are (a) $Mn_{0.5}Zn_{0.5}Fe_2O_x$, (b) $Mn_{0.24}Zn_{0.69}Fe_{2.1}O_x$, (c) $Mn_{0.65}Zn_{0.65}Fe_{1.7}O_x$ and (d) the target composition of $Mn_{0.7}Zn_{0.3}Fe_2O_x$. Open squares and closed triangles denote the samples annealed at a pressure of 1 x 10^{-4} Pa and in an air atmosphere, respectively.

References

[1] Miyanishi S, Iketani N, Takayama K, Innami K, Suzuki I, Kitazawa T, Ogimoto Y, Murakami Y, Kojima K and Takahashi A 2005 *IEEE Trans. Magn.* **41** 2817

[2] Williams M C, Chrisey B D, Lubits P, Grabowski S K and Cotell M 1994 *J. Appl. Phys.* **75** 1676

[3] Kumari S, Tripathi C, Singh P A, Chauhan D, Shrivastav R, Dass S and Satsangi R V 2006 *Curr. Sci.* **91** 1062

Journal of Physics: Conference Series **191** (2009) 012028 doi:10.1088/1742-6596/191/1/012028

Optical properties of Eu-implanted GaN and related-alloy semiconductors

A Nishikawa, H Kasai, T Kawasaki, Y Terai and Y Fujiwara
Division of Materials and Manufacturing Science, Graduate School of Engineering, Osaka University
2-1 Yamadaoka, Suita, Osaka 565-0871, Japan

E-mail: nishikawa@mat.eng.osaka-u.ac.jp

Abstract. We investigated the optical properties of Eu-implanted GaN and related-alloy semiconductors by photoluminescence (PL) and time resolved PL measurements at room temperature (RT). After thermal annealing, the red emission was observed in each sample due to the intra-4f transitions in Eu^{3+} ions. The PL intensity increases with Al incorporation into GaN while it decreases with In incorporation into GaN, indicating that the emission efficiency depends on the host materials while the emission wavelength is independent of the host materials. Since the decay times of Eu emission of the samples are found to be similar by time resolved PL measurements, the origin of the Eu emission of each sample is supposed to be same. Therefore, the difference in the PL intensity is attributed to the difference in the energy transfer efficiency because of the ion implantation damage and/or bandgap energy of the host materials.

1. Introduction

Rare-earth (RE) doped semiconductors have been intensively investigated due to their superior optical properties such as a very sharp emission from an intra-4f transition [1]. As a host material, GaN is a promising candidate because its wide-gap allows visible wavelength emission from RE ions [2]. The energy transfer mechanism from the host material to Eu^{3+} ions was proposed by Lee *et al.*, in which excitation energy is directly transferred from the host material to the $^5D_3$ state of the Eu ion through resonant energy transfer or nonresonant Auger scattering, or indirectly transferred through defect/trap level located at 0.29 eV below the conduction band of GaN to the $^5D_3$ state of the Eu ion by nearly resonant Auger scattering [3]. Therefore, the bandgap energy of the host materials would play an important role in the energy transfer efficiency from the host material to Eu ions. Thus, it is of great interest to investigate the optical properties of Eu-doped GaN and related-alloy semiconductors. In this contribution, Eu-ions were implanted into GaN, InGaN (In: 8, 13, 17%) and AlGaN (5, 12, 18%) to investigate the effect of the host material for RE doping.

2. Experimental procedure

Samples were grown on c-plane sapphire substrates by organometallic vapor phase epitaxy (OMVPE). The group III and V sources were trimethylgallium, trimethylaluminium, trimethylindium and ammonia. Figure 1 shows the structures of GaN, $Al_xGa_{1-x}N$, and $In_xGa_{1-x}N$ samples used in this study. The structure of the GaN sample consists of a 1.5 μm-thick GaN layer and low-temperature (LT)-grown GaN buffer layer, for InGaN or AlGaN samples. InGaN (50 nm) or AlGaN (50 ~ 120 nm)

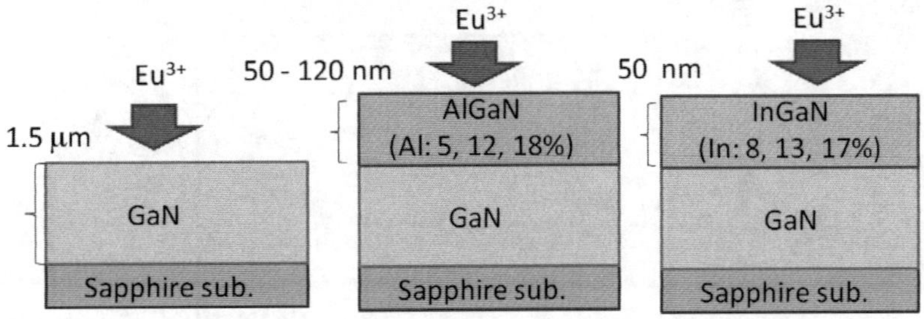

Figure 1. Sample structures of the GaN, AlGaN, and InGaN samples.

layers were grown on the 1.5 μm-thick GaN layer, respectively. Al or In incorporation into GaN were performed during OMVPE growth. The In and Al compositions were 8, 13 and 17% in InGaN layers and 5, 12 and 18% in AlGaN layers, respectively. In general, In incorporation into GaN tends to degrade the crystalline quality due to large lattice mismatch between GaN and InN. Eu ion implantation was carried out with an accelerating voltage of 200 keV. The dosage of the implanted Eu was 1×10^{14} cm^{-2}. According to SRIM2008 simulation, the projected range and peak concentration were around 40 nm and 5×10^{19} cm^{-3}, respectively. After ion implantation, the samples were annealed at 1000°C in nitrogen ambient for 1 min. Photoluminescence (PL) measurements were carried out at room temperature (RT) with a He-Cd laser (λ=325 nm) for excitation. Time resolved PL measurements were carried out at RT using a pulsed N_2 laser (λ=337 nm). Because of the limitation of the excitation wavelength, the decay times of PL emission for only GaN and InGaN samples were measured.

3. Results and Discussion

Figure 2 shows PL spectra of the Eu-implanted GaN, $In_{0.17}Ga_{0.83}N$, and $Al_{0.18}Ga_{0.82}N$ samples. The red emission was observed in each sample after thermal annealing, although no peak was observed in the observed spectral range from 350 to 700 nm just after ion implantation. Emission peaks observed at 543, 600, 622, 634 and 665 nm are associated with intra-4f shell transitions of $^5D_1$-$^7F_1$, $^5D_0$-$^7F_1$, $^5D_0$-$^7F_2$, $^5D_1$-$^7F_4$, and $^5D_0$-$^7F_3$ in Eu^{3+} ions, respectively [4, 5]. In addition to these peaks, several minor peaks were observed in this range, indicating the presence of Eu-ions in lower site symmetry than relaxed trigonal symmetry [6]. Indeed, X-ray diffraction measurements revealed the Eu-induced strain in the samples after ion implantation. While the emission wavelengths remain constant in spite of the host

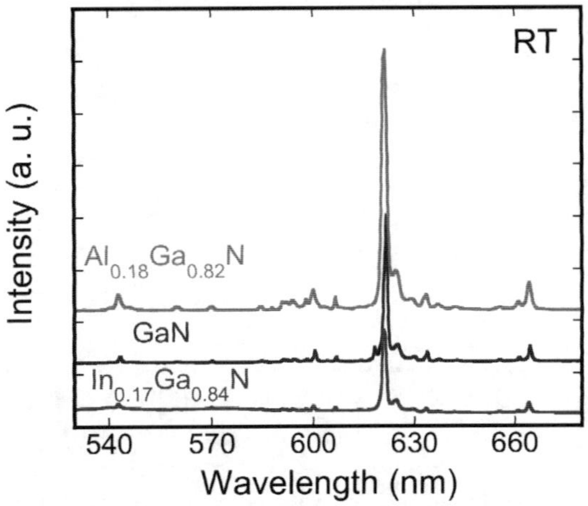

Figure 2. PL spectra of the samples after thermal annealing measured at RT.

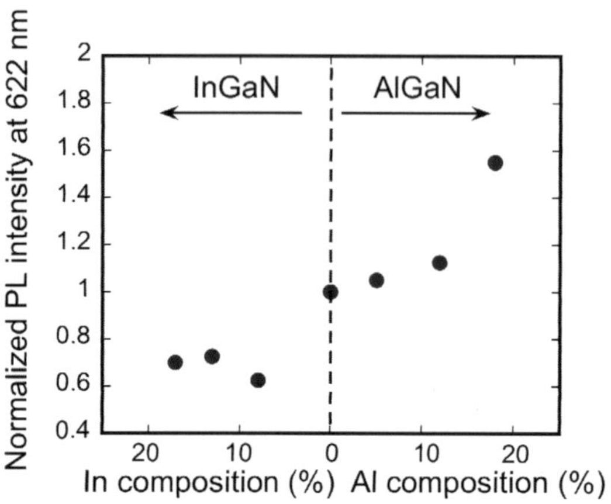

Figure 3. In and Al composition dependence of the PL peak intensity at 622 nm.

material difference, the PL intensity increases with Al incorporation into GaN and it decreases with In incorporation into GaN.

Figure 3 shows the In and Al composition dependence of the PL peak intensity at 622 nm. In incorporation degrades the PL intensity while Al incorporation enhances it. The increase in the emission intensity for Eu-implanted AlGaN is consistent with the report by Wakahara [7], in which it is attributed to the increase of energy transfer efficiency and/or the amount of active Eu ions. In the case of the Eu-implanted InGaN samples, what causes the degradation of PL intensity? Besides the bandgap of the host material, the local symmetry or strain associated with the host material and the crystal quality of the host material also affect the emission efficiency of Eu ions because the excitation energy for Eu ions are transferred from the host material.

In order to investigate the decay characteristics of the Eu emission at 622 nm, time resolved PL measurements were carried out at RT. Figure 4 shows the transient decay of the $^5D_0$-$^7F_2$ transition for the $In_{0.17}Ga_{0.83}N$ and GaN samples. The transient decay of PL intensity of both samples can be fitted by assuming double exponential feature with decay times of fast and slow components. For the $In_{0.17}Ga_{0.83}N$ sample, the fast and slow decay times were 34 and 165 μs, respectively, while they were 33 and 147 μs, respectively, for the GaN sample, which are consistent with the previous reports [3, 6].

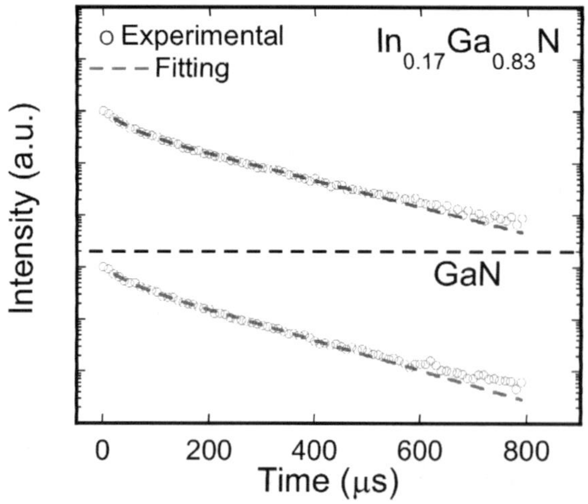

Figure 4. Transient decay of the $^5D_0$-$^7F_2$ transition for the $In_{0.17}Ga_{0.83}N$ and GaN samples.

In general, the Eu luminescence properties such as the emission wavelength and the decay time are influenced by the host material due to local strain [5]. However, the similarity of the decay times of the GaN and $In_{0.17}Ga_{0.83}N$ samples indicates that the origin of the Eu emission from the GaN and $In_{0.17}Ga_{0.83}N$ samples is supposed to be same. Therefore, the difference in the PL intensity is attributed to the difference in the energy transfer efficiency from the host material to Eu ions due to the ion implantation damage and/or the bandgap energy of the host materials, rather than the difference in the transition probability of Eu emission.

4. Conclusions

We investigated the optical properties of Eu-implanted GaN and related-alloy semiconductors. The red emission was observed in each sample with He-Cd laser excitation at RT. The PL intensity increases with Al incorporation into GaN, while it decreases with In incorporation into GaN. The decay times of Eu emission of the samples are found to be similar, indicating that the origin of the Eu emission is supposed to be same in spite of the difference in the host materials. Therefore, these results suggest that the energy transfer efficiency from the host material to Eu ions could be controlled by In or Al incorporation into GaN to change the bandgap energy of the host material, leading to the high luminescence efficiency from Eu ions.

Acknowledgments

This work was supported in part by a Grant-in-Aid for Creative Scientific Research No.19GS1209 from the Japan Society for the Promotion of Science. This work was also supported in part by a Grant-in-Aids for Exploratory Research No. 19656082, and the Global COE Program "Advanced Structural and Functional Materials Design" from the Ministry of Education, Culture, Sports, Science and Technology (MEXT), Japan.

References

[1] Ehenen H, Schneider J, Pomrenke G and Axmann A, 1983 *Appl. Phys. Lett.* **43** 943
[2] Heikenfeld J, Garter M, Lee D S, Birkhahn R and Steckl A J, 1999 *Appl. Phys. Lett.* **75** 1189
[3] Lee C W, Everitt H O, Lee D S, Steckl A J and Zavada J M, 2004 *J. Appl. Phys.* **95** 7717
[4] Dieke G H and Crosswhite H M, 1963 *Appl. Optics* **2** 675
[5] Monteiro T, Boemare C, Soares M J, Sa Fereira R A, Carlos L D, Lorenz K, Vianden R and Alves E, 2001 *Physica B* **308-310** 22
[6] Peng H, Lee C W, Everitt H O, Munasinghe C, Lee D S and Steckl A J, 2007 *J. Appl. Phys.* **102** 073520
[7] Wakahara A, 2006 *Opt. Mater.* **28** 731

Estimation of the Number of Cracks Generated by Compressive Stress in Polycrystalline Graphite using Ultrasonic Method

R Furushima[1,*], Y Matsuo[1], T Shiota[1], J. S. Cross[2] and K Yasuda[1]

[1] Department of Metallurgy and Ceramics Science, Tokyo Institute of Technology, Japan

[2] Graduate School of Science and Engineeing, Tokyo Institute of Technology, Japan

* Present Affiliation: Department of Materials Science and Technology, Nagaoka University of Technology, 1603-1 Kamitomioka, Nagaoka, Niigata, Japan

E-mail: fryoichi@mst.nagaokaut.ac.jp

Abstract. A theory was developed to estimate the change in the apparent sonic velocity in porous ceramics by solving Lavrov's integral equation of the cylindrical inclusion problem. It was shown that the change in apparent sonic velocity was caused by phase shift of the sonic wave. To extend the theory to the case of multiple inclusions, the superposition principle was applied. The theory was verified by analyzing the apparent sonic velocity during a compression test of polycrystalline graphite, and the number of cracks that formed during loading was estimated to be 700, when the dimensions of crack length and width were assumed to be 1mm and 10μm, respectively.

1. Introduction

Elastic inhomogeneity affects propagating waves in materials. This phenomenon has been utilized in the ultrasonic pulse echo method to detect the location of a crack [1]. However, it is difficult to use this method if a material contains multiple cracks. In addition, damage evolution itself is one of the most important factors in materials characterization now. In the ultrasonic pulse echo method, parameters such as apparent sonic velocity and attenuation coefficient have been used to estimate damage processes [2-4]. However, the quantitative relationship between these parameters and microstructural change (or micro damage process) has not been fully clarified yet. So, we need to analyze the relationship between the propagating wave change and the number of cracks for quantitative understanding of the damage processes.

Lavrov *et al.* studied the change in a plane elastic wave caused by a cylindrical inclusion [5]. This report suggests that it is possible to estimate the relationship between the wave change and the microstructural change in a material. To calculate the wave change due to multiple cracks, an assumption that the crack is a cylindrical inclusion must be introduced.

In this paper, we solved the wave equation suggested by Lavrov *et al.* for phase shift in a plane elastic wave due to a single cylindrical inclusion located in an isotropic elastic material. Then, the phase shift was superimposed into the frequency space in order to estimate the total phase shift due to multiple cracks. This value was converted into the change in apparent sonic velocity that was a parameter in the ultrasonic pulse echo method. From these results, we can calculate the number of cracks necessary to change the apparent sonic velocity.

The goal of this study is to propose a model for estimating the number of cracks generated in a material, that results in a change in the apparent sonic velocity. To verify this model, uni-axial compression tests were conducted in polycrystalline graphite, and the number of cracks was estimated using an arbitrary stress.

2. Theory of the wave change

We consider the system where plane elastic longitudinal wave propagates through an elastic matrix and impinges on a cylindrical inclusion. Figure 1 shows the configuration of the wave and the inclusion. The propagating direction of the plane wave is vertical to that of the longitudinal axis of the inclusion. The length and diameter of the inclusion are $2L$ and $2h$, respectively, assuming that L is much larger than h. Both the matrix and the inclusion are linear elastic bodies with density ρ, ρ_1, respectively.

After the wave passes through the inclusion, its direction is maintained if scattering of the wave is neglected by the inclusion. Therefore, we have one non-zero component of the displacement parallel to the x_3 axis in figure 1. This component can be calculated from an integral equation proposed by Lavrov et al. [6].

The relationship between the displacement u_3 in the material containing one inclusion and that u_0 without inclusion is formulated as below.

$$u_3 = R \cdot \exp(i\theta) \cdot u_0 \qquad (1),$$

,where R and θ is the wave amplification ratio and the phase shift due to the single inclusion, respectively. Equation (1) results from the wave change by propagating through the inclusion, where u_3 is the complex number. In Eq. (1), R and θ can be expressed as follows,

$$R = \left[\{\mathrm{Re}(u_3/u_0)\}^2 + \{\mathrm{Im}(u_3/u_0)\}^2\right]^{1/2} \qquad (2)$$

$$\theta = \tan^{-1}\left(\frac{\mathrm{Im}(u_3/u_0)}{\mathrm{Re}(u_3/u_0)}\right) \qquad (3),$$

where R and θ are both functions of frequency.

Generally, the incident wave has multiple frequencies. So, the pulse wave $x(t)$ is separated into waves with a single frequency by a Fourier series expansion as follows

$$x(t) = \frac{a_0}{2} + \sum_{n=1}^{N_{max}} \left[a_n \cos(2\pi f_n t) + b_n \sin(2\pi f_n t)\right] \qquad (4),$$

,where a_n and b_n denote Fourier expansion coefficients corresponding to cosine and sine waves with frequency f_n, respectively.

For the case where the pulse wave propagates through multiple inclusions, the calculations are simplified for K cylindrical inclusions by assuming vertical alignment to the wave propagation axis (the x_3

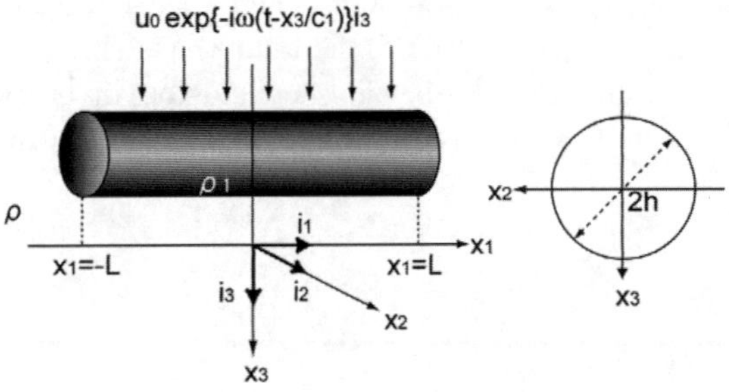

Figure 1 The configuration of the waves and the inclusion

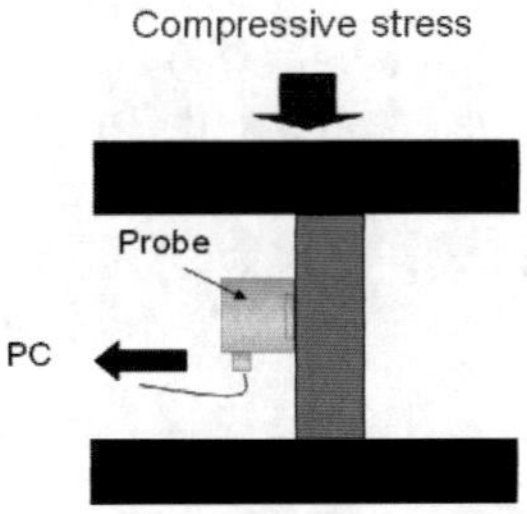

Figure 2 An illustration of the compression test

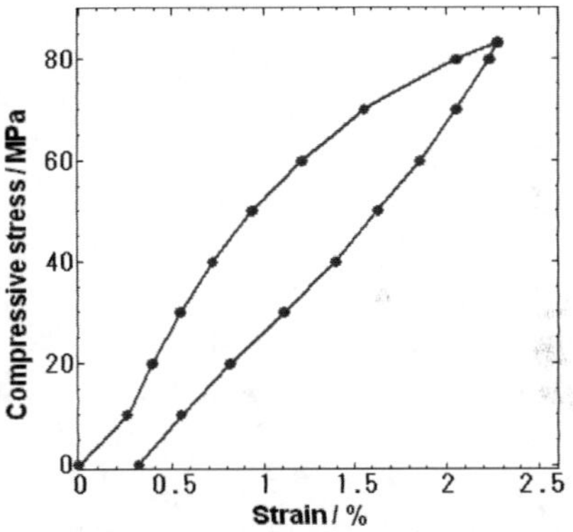

Figure 3 Stress-strain curve for a polycrystalline graphite specimen

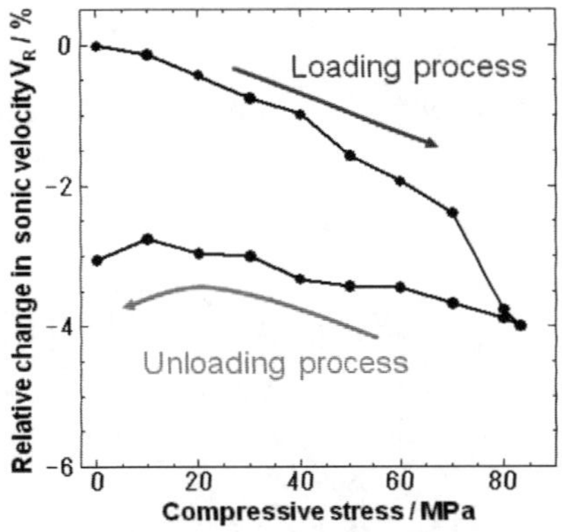

Figure 4 The change in apparent sonic velocity as a function of compressive stress

axis in figure 1) and each center of gravity is located along the x_3 axis. If the interactions among inclusions are neglected, the total phase shift by multiple inclusions is obtained by superimposing the phase shift due to the single inclusion in the frequency space. The output wave $y(t)$ after passing through these inclusions is obtained in the following form,

$$y(t) = \frac{a_0}{2} + \sum_{n=1}^{N_{\max}} (R_n)^K \{a_n \cos(2\pi f_n t + K\theta_n) + b_n \sin(2\pi f_n t + K\theta_n)\} \quad (5)$$

3. Experimental

Polycrystalline graphite (TOYO TANSO Co. Ldt.) specimens with dimensions of 30 mm in width (L_1), 30 mm in height (L_2) and 10 mm in thickness (L_3) were prepared for uni-axial compressive tests. Figure 2 shows an illustration of our experiments. The maximum applied stress was set to 83.3MPa, which is 90% of the mean fracture stress of the graphite. The loading axis was parallel to the direction of the height of the specimen and crosshead speed was 0.1 mm/min. During the loading-unloading process, ultrasonic measurements were carried out simultaneously. A transducer for transmitter-cum-receiver of longitudinal sonic waves (TOSHIBA TUNGALOY Co. Ldt.) was attached to the 30 × 30 mm face. The nominal frequency of the incident pulse wave was set to 1.0 MHz.

During the ultrasonic measurements, some echo peaks were observed. From the time interval of these peaks and thickness of the specimen, sonic velocity V_A was calculated. As an acoustic parameter for damage evaluation, we calculated the relative change in sonic velocity V_R as follows,

$$V_R \equiv \frac{V_A - V_{A0}}{V_{A0}} \quad (6),$$

where V_{A0} denotes sonic velocity before the first loading.

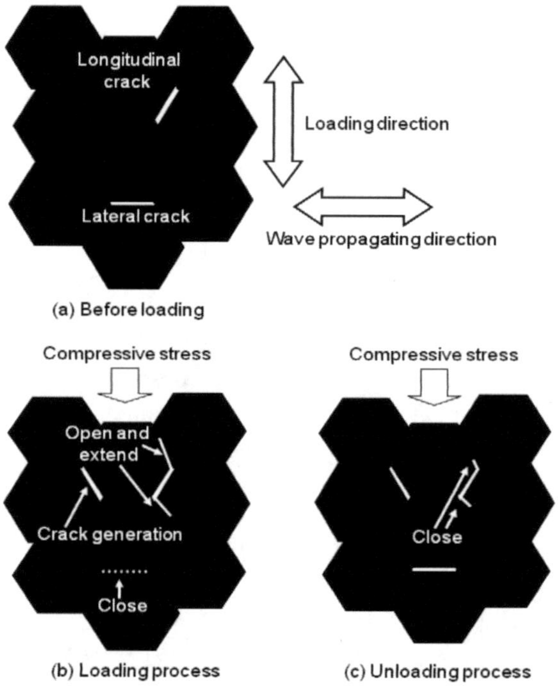

Figure 5 A damage model for different steps

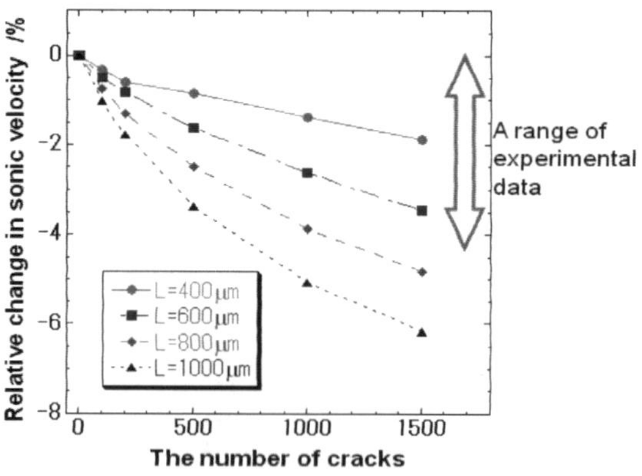

Figure 6 The relationship between relative change in sonic velocity V_R and the number of cracks (Crack length: L, Width: 0.01L)

4. Experimental results

Figure 3 shows a stress-strain curve and figure 4 shows the change in apparent sonic velocity as a function of compressive stress. In the loading process, the relative change in apparent sonic velocity V_R decreases with an increase in stress, which corresponds to the nonlinearity of the stress-strain curve. These results suggest that cracks were generated during the loading step, because both V_R and Young's modulus must decrease due to crack generation [7-8]. During the unloading step, V_R increases and then decreases with stress. Since the unloading curve has an inflection point at around 20 MPa, the curve changes from concave to convex. This change must be consistent with a slight decrease in V_R while unloading below 20 MPa in figure 4. In this region, a secondary damage process has taken place.

From the above results, we propose a damage model for polycrystalline graphite as shown in figure 5. We assume two types of cracks in a polycrystalline graphite specimen; one is aligned almost parallel to the loading axis and the other is aligned vertical to the loading axis. We define the former cracks as longitudinal cracks and the latter as lateral cracks as shown in figure 5 (a). In the loading process, new cracks form at the interface between graphite filler grains. These longitudinal cracks open and extend but lateral cracks close under the load. The opening and growth of cracks causes a decrease in V_R, and vice versa. In this case, the two effects are competing. However, the effect of longitudinal cracks dominates the change in V_R, because ultrasonic wave propagates in the direction vertical to the loading axis (viz. longitudinal cracks also). Therefore, V_R decreases during the loading process. For the unloading process, on the other hand, longitudinal cracks retract, and then lateral cracks begin to open again under 20 MPa. This mechanism contributes to the initial increase and the subsequent decrease in V_R in the unloading step.

5. Estimation of the number of cracks

Equation (5) provides a relationship between the number of cracks and the total phase shift in the pulse wave. Therefore, the number of cracks is obtained as the function of the change in V_R in figure. 6. This figure shows that the number of cracks is almost proportional to the relative change in sonic velocity in each case. If the decrease in V_R during the loading process is caused by an increase in the number of cracks, the increase in V_R during the unloading process must attribute to the decrease in the length of cracks. Quantitatively, the aspect ratio of the crack is assumed to be constant (L/h=100), and we propose that the damage of the specimen must be explained by the generation of 700 cracks with a length of 1 mm and width of 10 μm during the loading step. If a smaller crack size is selected, then the number of cracks increases. After unloading, the cracks may be partly closed, and their length and width are estimated to be 800 μm and 8 μm to fit the experimental data. These changes need to be verified by direct measurement of cracks, however, up to now, this has yet to be undertaken. This topic will require further work.

6. Conclusions

In this paper, a model is proposed for estimating the number of cracks generated by compressive loading of graphite. The total phase shift of the incident wave was an important factor to determine the velocity change by crack generation. Based upon this theoretical framework, the number of cracks was estimated in a polycrystalline graphite specimen subjected to uni-axial compressive loading. The damage model assumes that crack generation only occurs during loading, whereas, the longitudinal cracks close while unloading.

References

[1] Chiou C.-P., Schmerr L.W. 1991 *Ultrasonics,* **29** 471
[2] Mirmiran A. and Wei Y. 2001 *J. Engineering Mechanics*, **127** 126
[3] Niyogi S.K. and Das A.C. 1994 *InterCeram: International Ceramic Review,* **43** 453
[4] Kline R. A. and Hashemi D. 1985 *Ultrasonics Symposium Proceedings,* 1032
[5] N.A. Lavrov and E.E. Pavrovskaia 2001 *J. Sound and Vibration,* **248** 329
[6] Furushima R., Matsuo Y., Shiota T., Yasuda K. *Journal of Nondestructive Evaluation* (Submitted)
[7] Furushima R., Matsuo Y., Shiota T. and Yasuda K. 2007 *Journal of Materials Science*, **42** 8652
[8] Camus G., Guillaumat L. and Baste S. 1996 *Composites Science and Technology,* **56** 1363

Effect of Ultraviolet Light Irradiation on Etching Process of Poly(methyl methacrylate) by Ion Beam Injections

S Yoshimura[1], K Ikuse[1], Y Tsukazaki[1], M Kiuchi[1,2] and S Hamaguchi[1]

[1]Center for Atomic and Molecular Technologies, Graduate School of Engineering, Osaka University, 2-1 Yamadaoka, Suita, Osaka 565-0871, Japan

[2]National Institute of Advanced Industrial Science and Technologies (AIST), 1-8-31 Midorigaoka, Ikeda, Osaka 563-8677, Japan

E-mail: yosimura@ppl.eng.osaka-u.ac.jp

Abstract. Effects of ultraviolet (UV) light irradiation on polymer dry etching processes have been investigated with the use of a low-energy mass-selected ion beam injection system. Etching yields of poly(methyl methacrylate) (PMMA) by Ar or CF_3 ion beam injections were evaluated from the loss of PMMA film weight measured by a quartz crystal microbalance (QCM) during the ion beam injection process with or without simultaneous UV light irradiation. Significant enhancement of the etching yield of PMMA was observed during the simultaneous irradiation with CF_3 ion beam and UV light over the sum of its sputtering yields obtained from separate ion-beam or UV-light injection processes. By contrast, no significant change of etching yields was observed when the UV light was superposed on the Ar ion beam.

1. Introduction

Carbon tetrafluoride (CF_4) gases are widely used in semiconductor manufacturing processes such as etching of silicon dioxide (SiO_2) substrates. Etching yields of various materials due to fragment ions produced in CF_4 gas-based plasmas provide useful information on the control of manufacturing processes. Recently detailed measurements of ion energy dependence of SiO_2 etching yields by F^+, CF^+, CF_2^+, and CF_3^+ ions with an ion beam system have been reported [1-3].

While physical/chemical sputtering by ions plays an important role in CF_4 gas-based plasma processes, ultraviolet (UV) light may also affect etching processes because it can cause excitation of substrate atoms and change their average bond energies [4-5]. Although it is important to evaluate the effect of UV light on the etching, there had been no experiments to investigate the effect of light on the etching by F^+, CF^+, CF_2^+, and CF_3^+ ions. Recently we have measured SiO_2 etching yields by CF_3^+ ions with or without the UV light irradiation. No significant effect of the UV light on etching yields of SiO_2 has been observed in our experimental condition [6].

In this work, we have examined etching processes of poly(methyl methacrylate) (PMMA) by Ar^+ or CF_3^+ ions. PMMA was used as a simple model for various polymer materials (such as photoresist) used in SiO_2 etching processes. The typical atomic structure of PMMA is shown in figure 1. Photon wavelengths required to break atomic bonds in PMMA can be estimated to be similar to the bond energies approximately (340 nm for a C-C bond, 330 nm for a C-O bond and 290 nm for a C-H bond).

Journal of Physics: Conference Series **191** (2009) 012030 doi:10.1088/1742-6596/191/1/012030

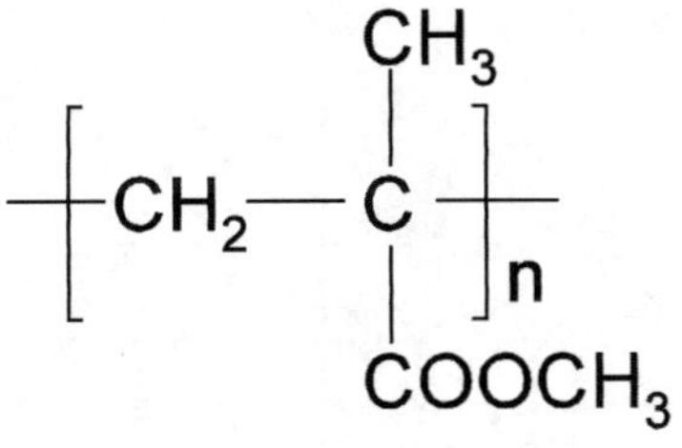

Figure 1. A diagram of the structural formula of poly(methyl methacrylate).

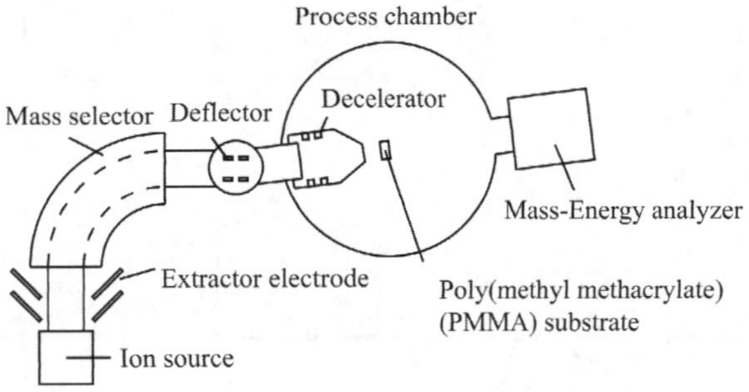

Figure 2. A schematic drawing of the low-energy mass-selected ion beam system (top view).

Thus PMMA is more likely to be affected by UV light if it is exposed to simultaneous irradiation with ions and UV light due to its relatively weak bonds compared to "harder" materials such as Si and SiO_2.

2. Experimental Apparatus

The experiment was carried out in a low-energy mass-selected ion beam system [7]. This system consists of a Freeman-type ion source, an extractor electrode, a magnetic-field-based mass selector, a deflector, a decelerator, a process chamber, and a mass-energy analyzer, as shown in figure 2. Ions produced from the source gas (Ar gas or Ar-diluted CF_4 gas in the present experiments) in the ion source chamber are extracted by a high voltage of -25kV applied to the extractor electrode. The ions are separated by the mass selector, so that only desired ion species (Ar^+ ions or CF_3^+ ions in the present experiments) remain in the beam line without impurity ions. The mass-selected ion beam is then deflected by a small angle, the process of which removes high energy neutral species generated by charge exchange reactions with residual gases in the beam line. Finally, the ion beam is decelerated to the desired kinetic energy and injected to the PMMA substrate. The injection angle is normal to the substrate. The quality of the ion beam (i.e., energy and mass distribution of incident ions) can be examined by the mass-energy analyzer (balzers, PPM-421), which consists of a cylindrical mirror analyzer for the energy analysis and a quadrupole mass spectrometer for the mass analysis.

The ion beam current was measured by the Faraday cup located in the process chamber with an ammeter (KEITHLEY, 6485). The number of incident ions can be evaluated from the ion beam current. Decrease of mass of the substrate material, i.e., PMMA in the present experiments, was evaluated from the change of resonance frequency of a quartz crystal microbalance (QCM, ULVAC, CRTM-9000). The QCM crystal is gold-coated AT-cut crystal (ULVAC, CRTS-0) and it has a resonance frequency of 5 MHz. Spin-coated PMMA (weight-average molecular weight : 100000) on a QCM crystal surface was used as a substrate. The PMMA substrate was set at room temperature. The details of our QCM system are described in [8]. The QCM was located just below the Faraday cup on the manipulator in the process chamber. By moving the manipulator vertically, one can select ion beam injection into either the Faraday cup or the substrate. When the ion beam is injected into the mass-energy analyzer, the manipulator is pulled up to the uppermost position.

The PMMA substrate surface was irradiated with UV light at an oblique angle (50 degree), as shown in figure 3. In the present experiments, Ar inductively coupled plasma (ICP) was used as the UV light source. The base pressure in the Pyrex chamber (50 mm in diameter, 350 mm in length) of ICP device was about 0.4 Pa and the pressure of Ar gas was about 4 Pa. The Ar plasma was produced by the 13.56 MHz power source (NIHON KOSHUHA, HFS-005) at 30 W. The distance between the ICP device and the PMMA substrate was about 40 cm. The Ar ICP light was set to pass through an MgF_2 window, which allowed light with the wavelength of 110 nm or longer. The Ar ICP light seems to have a wide spectrum including UV and visible light. Figure 4 shows an emission spectrum of the

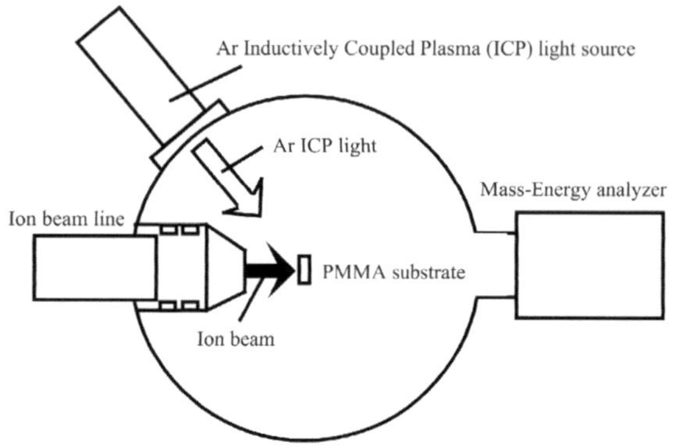

Figure 3. A schematic drawing of the process chamber (top view).

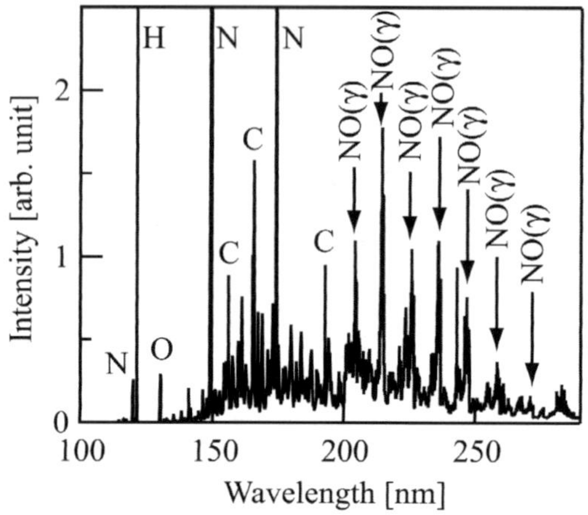

Figure 4. The vacuum ultraviolet emission spectrum of an inductively coupled plasma produced from Ar gas.

Ar ICP measured by a Seya-Namioka monochromator in the vacuum ultraviolet region. Line spectra of impurities (hydrogen H, carbon C, nitrogen N, and oxygen O) are seen in the 110-200 nm region while the emission is dominated by NO(γ) bands in the 200-290 nm region [9]. It is seen that there are no dominant line spectra of Ar in this region (110-290 nm). The UV light intensity measured by a UV power meter (HAMAMATSU, H8025-146, sensitive range = 120-200 nm) were found to be 8 $\mu W/cm^2$.

Etching yields of the PMMA film were determined from the loss of its weight measured by the QCM and the ion beam current measured by the Faraday cup. In the experiments presented here, the etching yields due to ion injections only were compared with those due to simultaneous ion and UV light irradiation. The typical measurement procedures are as follows. First a time-invariant (i.e., steady-state) ion beam is injected into the Faraday cup for the beam current measurement. After the exact value of the beam current is determined, the same ion beam is then injected into a substrate film deposited on a QCM, which measures the amount of etching during the process. In the case of simultaneous injections mentioned above, the surface is also irradiated with UV light (figure 3). At the end of etching experiments, the ion beam is again measured by the Faraday cup for the confirmation of steadiness of the beam current during the experiment.

3. Experimental Results

First we investigated the etching of PMMA by the Ar ion beams. Figure 5 shows the normalized etched depth of the PMMA substrate over 20 minutes as a function of beam duration time. The normalized etched depth, i.e., the decrease of the PMMA film thickness per unit ion beam current, is proportional to the etching yield. The ion beam energy was set at 300 eV. The Ar ion beam current was about 1 μA. The power of Ar ion beam was about 600 $\mu W/cm^2$ and was larger than the power of UV light (8 $\mu W/cm^2$). Until 60 minutes, the Ar ion beam was injected to the PMMA substrate without the light irradiation from the Ar ICP. Then, while the Ar ion beam continued to be injected, the Ar ICP was turned on at t = 80 min. Therefore, the substrate was subject to the simultaneous irradiation with the Ar ion beam and the Ar ICP light from 80 to 180 minutes. It is seen that no significant change in etched depth is observed before and during the Ar ICP light irradiation. At t = 200 min, the light source was again turned off and, after this time, only the Ar ion beam was injected again.

Second we investigated the etching of PMMA by the CF_3 ion beams (figure 6). The energy of CF_3 ions was set to be 470 eV. The CF_3 ion beam current was about 0.1 μA. The power of ion beam was about 100 $\mu W/cm^2$. The Ar ICP light was irradiated from 80 to 140 minutes. As clearly seen in figure

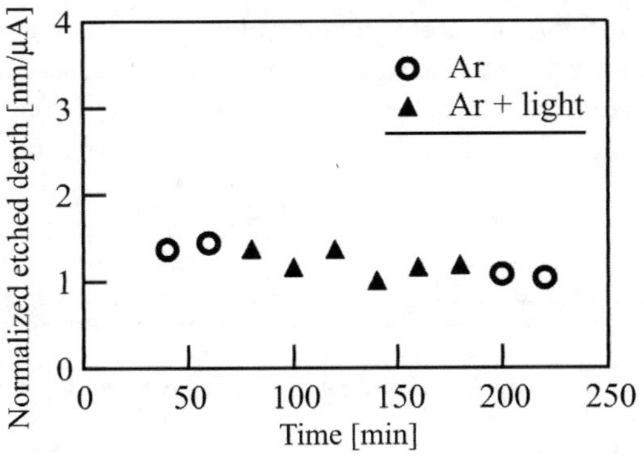

Figure 5. The etched depth normalized by the ion beam current for duration of 20 min of the PMMA substrate during the Ar ion beam injection at 300 eV. From 80 to 180 min, light from the Ar ICP light source was also irradiated. The open circles show the etched depths by the Ar ion beam only and the closed triangles correspond to those by the simultaneous irradiation with the ion beam and light from the Ar ICP light source.

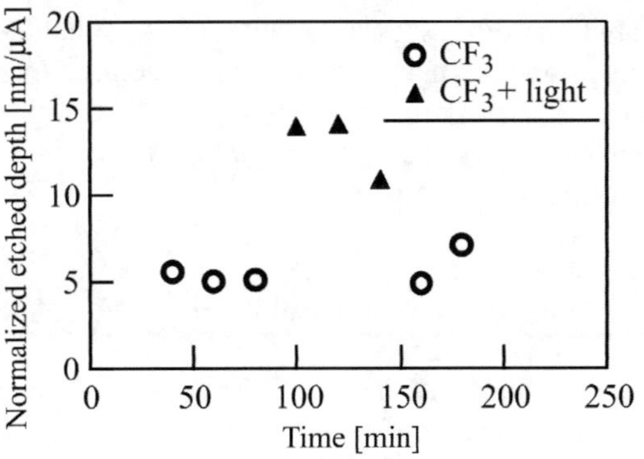

Figure 6. The etched depth normalized by the ion beam current for duration of 20 min of the PMMA substrate during the CF_3 ion beam injection at 470 eV. From 100 to 140 min, light from the Ar ICP light source was also irradiated. The open circles show the etched depths by the CF_3 ion beam only and the closed triangles correspond to those by the simultaneous irradiation with the ion beam and light from the Ar ICP light source.

6, the etched depth increased significantly – up to almost 3 times – during the simultaneous irradiation with the ion beam and the Ar ICP light. It should be also noted that the etching yield went back to almost the same value as the original one after the termination of the Ar ICP light. In terms of the quantity of the "etched depth" given in figure 6, its average value in the case of the ion beam injection only was about 0.50 nm (for 20 minutes) and that in the case of the simultaneous injection of the ion beam and the Ar ICP light was about 1.51 nm (for 20 minutes). For comparison, in the case of the Ar ICP light irradiation only, it was only about 0.38 nm. These results show that significant enhancement of the etching yield of PMMA was observed during the simultaneous irradiation with CF_3 ion beams and UV light over the sum of its sputtering yields obtained from separate ion-beam or UV-light injection processes. The increase of the measured sputtering yield is likely to be synergetic effects of the ion beam and the UV light.

After the injections of the Ar ions, the CF_3 ions, or the UV light, no significant change of the PMMA substrates was observed at a glance. Now, we are trying to analyze the ion or UV injected PMMA substrates by the Raman spectroscopic analysis and the infrared spectroscopic analysis in order to investigate the change of C-C, C-O, and C-H bonds by the ion and UV injections.

4. Summary

In this work, we have evaluated the etching rates of PMMA by Ar or CF_3 ion beams with or without UV irradiation. The etched depths were compared among the three cases: (1) ion beam injections only, (2) simultaneous irradiation with an ion beam and light from an Ar ICP, and (3) irradiation with light from the Ar ICP only. As shown in figure 6, the etched depth increased significantly when the PMMA substrate was irradiated by both the CF_3 ion beam and the light simultaneously. The increased etched depth was much higher than the simple sum of the etched depths that could be obtained from separate irradiation with the CF_3 ion beam or the Ar ICP light. When the light source was turned off and only the CF_3 ion beam injection continued, then the value of etched depth went back to its original value, i.e., the value in the case of the CF_3 ion beam injection only. This result indicates that the increased etching yield during the simultaneous irradiation was caused not by the alteration (degeneration) of the PMMA substrate due to UV irradiation but most likely by some synergetic effects of the simultaneous

Journal of Physics: Conference Series **191** (2009) 012030 doi:10.1088/1742-6596/191/1/012030

irradiation. For example, excitation of substrate materials by UV light may result in weakening of atomic bonds temporarily and ion injections into the materials under such conditions may increase the etching yield significantly. By contrast, no significant change of etched depth was observed in the case of simultaneous irradiation with the Ar ion beams and the Ar ICP light in their energy and flux ranges examined here.

References

[1] Ishikawa K, Karahashi K, Tsuboi H, Yanai K and Nakamura M 2003 *J. Vac. Sci. Technol.* A **21** L1

[2] Karahashi K, Yanai K, Ishikawa K, Tsuboi H, Kurihara K and Nakamura M 2004 *J. Vac. Sci. Technol.* A **22** 1166

[3] Yanai K, Karahashi K, Ishikawa K and Nakamura M 2005 *J. Appl. Phys.* **97** 053302

[4] Ishikawa K, Okigawa M, Ishikawa Y, Samukawa S and Yamasaki S 2005 *Appl. Phys. Lett.* **86** 264104

[5] Samukawa S, Jinnai B, Oda F and Morimoto Y 2007 *Jpn. J. Appl. Phys.* **46** L64

[6] Ikuse K, Yoshimura S, Takizawa T, Karahashi K, Kiuchi M and Hamaguchi S 2008 *J. Vac. Soc. Jpn.* **51** 158

[7] Yoshimura S, Toh A, Sugimoto S, Kiuchi M and Hamaguchi S 2006 *Jpn. J. Appl. Phys.* **45** 8204

[8] Hine K, Yoshimura S, Ikuse K, Kiuchi M, Hashimoto J, Terauchi M, Nishitani M and Hamaguchi S 2008 *Thin Solid Films* **517** 835

[9] Rahman A, Yalin A P, Surla V, Stan O, Hoshimiya K, Yu Z, Littlefield E and Collins G J 2004 *Plasma Sources Sci. Technol.* **13** 537

Journal of Physics: Conference Series **191** (2009) 012031 doi:10.1088/1742-6596/191/1/012031

Roughness Increase on Surface and Interface of SiO_2 Grown on Atomically Flat Si (111) Terrace

K Ohsawa[1], Y Hayashi[1], R Hasunuma[1, 2] and K Yamabe[1, 2]
[1]Inst. Of Appl. Phys. Univ. of Tsukuba
[2]Tsukuba Research Center for Interdisciplinary Material Science, Univ. of Tsukuba

E-mail: hasunuma@bk.tsukuba.ac.jp

Abstract. With analysis on the surface and the interface roughness of SiO_2 films, it was found that roughness increased on both surface and interface only in the initial oxidation stage, where the oxide is thinner than 10 nm. On the other hand, for the thicker region, the roughness increase tends to saturate. By using atomically flat Si (111) terrace, such roughness increases of the surface and the interface can be observed. In addition, the thickness of 10 nm was found to be a threshold at which the mode of oxide growth turns to the linear growth from the initially enhanced growth. By taking account of Si emission at the SiO_2/Si interface, a model has been demonstrated to explain surface roughening of SiO_2.

1. Introduction

The integrated circuits have been developed with scaling down of electronic devices such as metal-oxide-semiconductor field effect transistors (MOSFETs). The characterization of SiO_2 surfaces and SiO_2/Si interfaces is becoming more essential under the scaling trend. Recently, the thickness of the gate oxide films is a few nm. For such thin oxide films, the leakage current is not negligible because of quantum effect of electron. Furthermore, even atomic scale roughness at the oxide surface and interface would deteriorate the uniformity of the oxide thickness [1]. The nonuniformity would cause the dispersion of the leakage current, and lowers production yield. Therefore, reduction of the roughness at the oxide surface and interface is considered as effective to achieve the uniform oxide. However, there still exists a problem that the oxidation mechanism has not been completely understood because of surface micro roughness at a mirror-polished Si wafer, although formation of the atomic scale roughness should be strongly related with the oxidation manner.

The oxidation process has been well explained by Deal-Grove theory, in which the oxidation proceeds with diffusion of oxidants in SiO_2 films and reaction between oxidants and Si at the interface [2]. For the very initial oxidation stage, however, it is widely known that this model can not explain the unexpectedly fast oxide growth rate when using dry O_2. To well explain the initial oxidation stage, Si emission model has been reported recently [3-5]. In this model it has been discussed that Si atoms are emitted at the SiO_2/Si interface during oxidation. When the oxide film is thin, the Si atoms diffuse through the oxide to the surface, and they desorb into the ambient as SiO molecules or encounter the oxidants followed by oxidation in the oxide film. This model has succeeded to explain the initial fast oxide growth described above. In this paper, by focusing on emitted Si atoms, we will discuss the mechanism of surface roughening.

2. Experimental

The mirror-polished p-Si (111) substrates were RCA cleaned. Then the samples proceeded to sacrificial oxidation at 1150 °C. After removing the sacrificial oxides in diluted hydrogen fluoride (HF), the atomically flat wide terraces were formed on the surfaces by wet etching in ultra low dissolved oxygen water (LOW) [6]. Finally, the samples were oxidized in 3.8% oxygen ambient diluted by Ar at 1000°C. The oxide thickness was varied from 1.5 to 40 nm. The thickness was measured by ellipsometry with the fixed refractive index of 1.46. The morphology of the SiO_2 surfaces was investigated with an atomic force microscope (AFM; SII Nano Technology Inc. SPI3800N+SPA-300HV) in cyclic mode in air. The SiO_2/Si interfaces were also characterized immediately after removing the oxide films by HF.

3. Results and Discussion

Figure 1 is the 200 nm x 200 nm AFM images of (a-c) SiO_2 surfaces, and (d-f) SiO_2/Si interface obtained after removal of the SiO_2 films. The oxide thickness is (a, d) 1.53, (b, e) 5.33, and (c, f) 13.1 nm, respectively. It is clear that the roughness on both surface and interface grows containing protuberances and concaves as the thickness increases. The increase in the interface roughness probably reflects two-dimensional oxide growth [7, 8]. Figure 2 summarizes the RMS values of the SiO_2 surfaces and the SiO_2/Si interface with variation of oxide thickness. Although the lateral resolution of our AFM is estimated as in the range of several 10 nm, the RMS values can give us relative roughness variation during oxidation, since the resolution in z-axis is about 0.01 nm. The plot in this figure represents average of the values obtained at three different points on a surface or interface. It should be noted that fluctuation of the three RMS values are negligible. It is interesting that roughness increase on both surface and interface are visible only in the initial oxidation stage, where the oxide is thinner than 10 nm. In addition, the RMS values for surfaces are saturated for the oxide thicker than 10 nm at around 0.12 nm. On the other hand, the interface RMS values oscillate around 0.1 nm. This oscillation seems to be caused by two-dimensional oxidation at the interface [7]. When there are many small SiO_2 islands at the interface, the interface RMS values should be large. When the islands grow and merge with each other, the RMS values should be small. Taking account

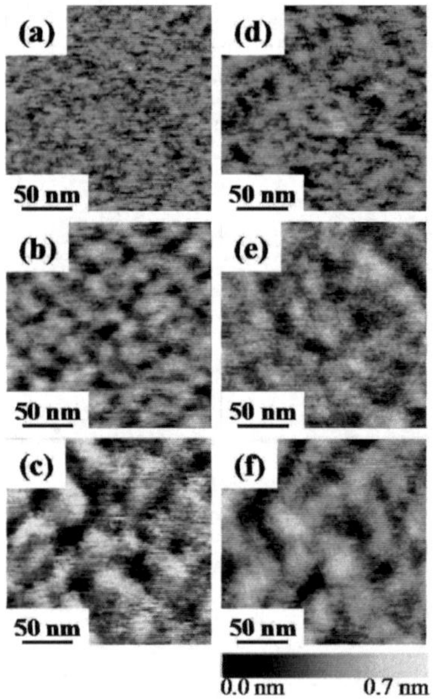

Fig. 1 200 nm x 200 nm AFM images of (a-c) SiO_2 surfaces, and (d-f) SiO_2/Si interface obtained after removal of the SiO_2 films. The oxide thickness is (a, d) 1.53, (b, e) 5.33, and (c, f) 13.1 nm, respectively.

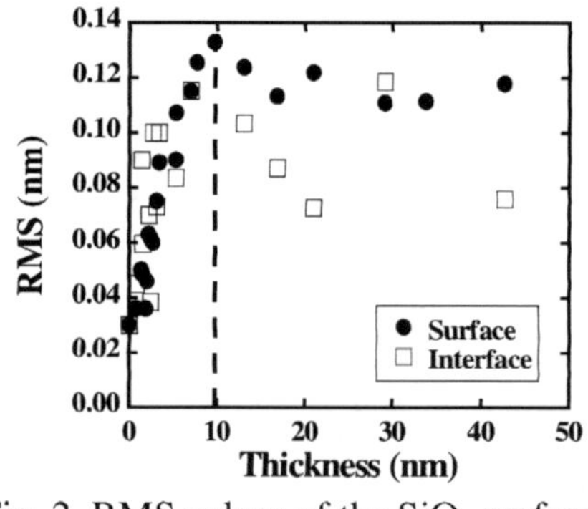

Fig. 2 RMS values of the SiO_2 surfaces and the SiO_2/Si interfaces.

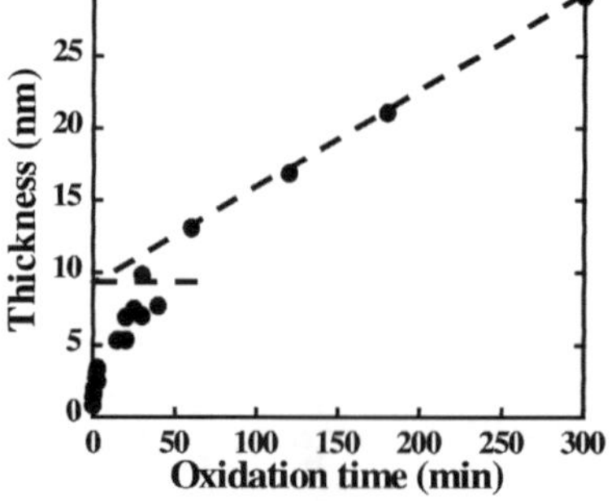

Fig. 3 Oxidation rate.

of typical RMS value of mirror-polished Si surface, which is larger than 0.1 nm, it should be emphasized that such roughness change can be observed only by applying the atomically flat Si surface.

Figure 3 shows the oxidation rate obtained for the present condition. In addition to linear growth mode which is well known as interface reaction-limited oxidation, the initial enhancement of the growth rate is also seen [2]. The initial growth rate has been explained by Si atoms emitting to reduce the stress at the oxide interface during oxidation [3]. Surprisingly, the thickness of 10 nm also seems to be a separation point for the two oxidation modes. This clearly indicates importance of consideration on oxidation mechanism to understand the roughness growth. It is quite reasonable to assume that emitted Si atoms contribute to roughness growth on SiO_2 surface, since when the emitted Si atoms are oxidized in the oxide film there should appear volume expansion which could increase the surface roughness. Figure 4 shows an illustration explaining surface roughness growth by taking account of Si emission. When the oxide is very thin, as illustrated in Fig. 4 (a), the emitted Si atoms can easily diffuse through the oxide out to the ambient in the form of SiO. In this case, the interface stress due to oxidation can be easily released by Si emission, which gives rapid oxidation. As the oxide thickness increases, some of the Si atoms should be oxidized by O_2 in the film near the surface, as illustrated in Fig. 4 (b). Oxidation of a single SiO molecule on or near the surface could induce protrusion of about 0.06-0.4 nm depending on how the volume expansion occurs. Thus, this oxidation near the surface could drastically increase the roughness on the surface. In addition, the gradient of Si concentration in the SiO_2 films decreases as the oxide thickness increases, and the Si atoms become hard to diffuse. The decrease of the gradient prevents the Si atoms from emitting and the stress at the interface would be accumulated. Therefore, the oxidation rate should be also decreasing. As oxide grows further, emitted Si atoms hardly diffuse out to the surface since most of them turn out to be oxidized in the film far from the surface (Fig. 4 (c)). In this case, the surface roughness would not increase because SiO_2 films could absorb the volume expansion in an isotropic manner. If we assume that there are sufficient O_2 in the thick SiO_2 film, all of the emitted Si atoms should be oxidized in the oxide before they are emitted out to the ambient. When the SiO_2 film is thick enough, the distance of the Si oxidized point from the interface can be considered as constant independent of the thickness, which gives rise to the constant gradient of Si concentration in the SiO_2 films. Consequently, the rate of Si emission becomes constant, and the oxidation proceeds in a linear growth manner. In this model, the surface roughness would increase only when the emitted Si atoms are oxidized in the film near the surface, and it is a good explanation for the results, as shown in Fig. 2, that the surface roughness increases only when the thickness is less than 10 nm. This model also well explains the relation between surface roughness and oxidation rate, and is consistent with that the thickness of 10 nm was found as the separation point in both.

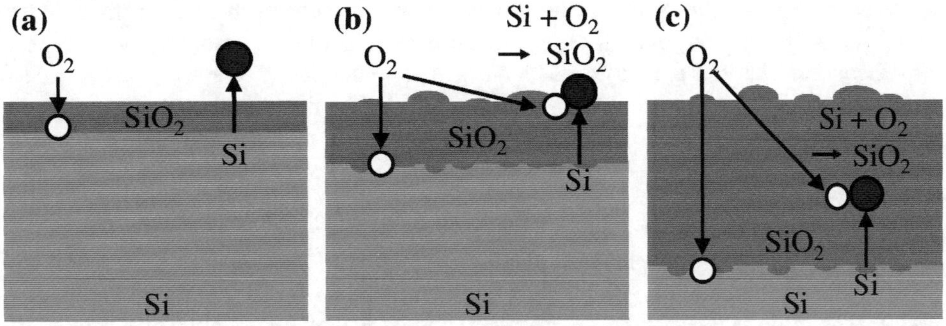

Fig. 4 The roughening model on SiO_2 surface taken account of Si emission when the oxide thickness is (a) very thin, (b) thin, and (c) thick. The open circles are O_2 molecules, and the filled circles are Si atoms.

However, to sufficiently explain the roughness growth with this model, we still have to discuss the possibilities of two-dimensional fluctuation of the phenomena described in the model. Specifically, it should be assumed that the emitted Si is oxidized at preferred sites in the film, or that Si emission occurs in a nonuniform manner at the interface. Figure 5 shows number of the protuberances on the surface and at the interface. It was found that, when the film is thinner than 10 nm, these values decreased as the thickness increases. In addition, the numbers remain unchanged in the region where the film is thicker than 10 nm. Figure 6 shows the average diameters of the protuberances on the surface and at the interface. When the oxide film is thinner than 10 nm, these grow up to roughly 10 nm with increasing the thickness. On the other hand, these remain at 10 nm in the region where the film is thicker than 10 nm. The most important feature is that two-dimensional growth of the protuberances on the oxide surfaces is similar to the morphological change at the interface. The results indicate that the interface roughness seems to define the surface roughness. The fact that the there exits two-dimensional roughness at the interface could lead us to assume that the emission of Si is nonuniform at the interface. This assumption is not questionable since it is known that oxidation at interface proceeds basically in a layer-by-layer manner with two-dimensional island formation [7, 8]. It can be easily imagined that, when the islands merge together, there exist intensive compressive stress around the boundaries in Si substrate. This stress would induce pronounced Si emission from the boundaries, which gives rise to nonuniform distribution of emitted Si atoms in the films. In consequence, oxidation of the Si atoms would be two-dimensionally nonuniform, with which the surface would locally expand resulting increase of RMS value.

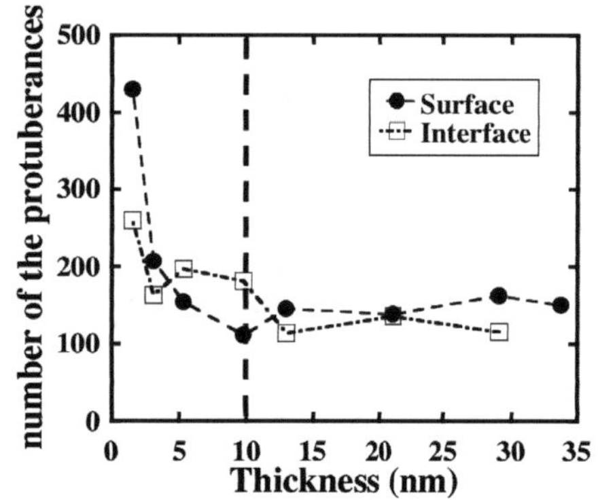

Fig. 5 The number of the protuberances on the SiO_2 surface and at the SiO_2/Si interface and the SiO_2/Si interfaces.

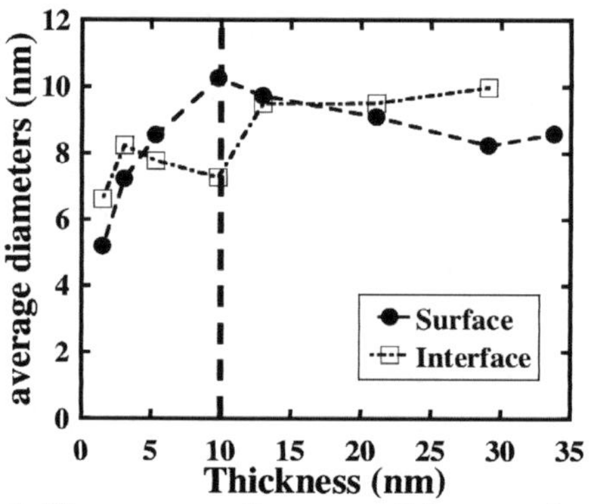

Fig. 6 The average diameter of the protuberances on the SiO_2 surface and at the SiO_2/Si interface and the SiO_2/Si interfaces.

4. Summery

With analysis on the morphology variation, it was revealed that both SiO_2 surface and interface were roughened during oxidation, even though the films were grown on atomically flat substrates. It directly brings a concern that there always exists a thickness fluctuation in atomic scale, especially for ultrathin oxides. It can be supposed that the thickness nonuniformity could have a significant impact on defining fluctuation in the electrical properties and reliability of nanometer scale devices.

References

[1] R. Hasunuma, J. Okamoto, N. Tokuda, K. Yamabe, Jpn. J. Appl. Phys. **43** (2004) 7861
[2] B. E. Deal and A. S. Grove, J. Appl. Phys. **36** (1965) 3770
[3] Y. Takakuwa, M. Nihei, and N. Miyamoto, Appl. Surf. Phys. **117/118** (1997) 141
[4] S. Ogawa, A. Yoshigoe, S. Ishidzuka, Y. Teraoka, and Y. Takakuwa, Thin Solid Films **508** (2006) 169
[5] M. Uematsu, H. Kageshima, and K. Shiraishi, Jpn. J. Appl. Phys. **39** (2000) 699
[6] N. Tokuda, D. Hojo, S. Yamasaki, K. Miki, and K. Yamabe, Jpn. J. Appl. Phys. **42** (2003) L1210
[7] D. Hojo, H. Oeda, N. Tokuda, and K. Yamabe, Jpn. J. Appl. Phys. **42** (2003) 1903
[8] T. Komeda, K. Namba, and Y. Nishioka, Jpn. J. Appl. Phys. **37** (1998) L214

Third International Symposium on Atomic Technology IOP Publishing
Journal of Physics: Conference Series **191** (2009) 012032 doi:10.1088/1742-6596/191/1/012032

Reliability Control of Silicon Dioxide Films by Radical and Thermal Oxidation

Z Lu[1], S Sato[1], R Hasunuma[1,2] and K Yamabe[1,2]
[1]Inst. of Appl. Phys. Univ. of Tsukuba
[2]Tsukuba Research Center for Interdisciplinary Materials Science, University of Tsukuba

Email: zhaolu@edu.esys.tsukuba.ac.jp

Abstract. Dielectric character of silicon dioxide formed by the combination of radical and thermal oxidation was studied. Constant current stresses were applied to study the degradation rate and breakdown lifetime of the oxides. Besides electric studies, AFM (Atomic Force Microscope) observation was also applied to see the roughness of SiO_2/Si surface and interface. It was revealed that the oxides formed by both radical and thermal oxidation inherited the advantage of the degradation rate of the thermal oxides and the advantage of flatness of SiO_2/Si interface of the radical oxides. We conclude that appropriate combination of oxidation methods can improve both the degradation of oxides and breakdown lifetime.

1. Introduction

MOSFET (Metal Oxide Semiconductor Field Effect Transistor) is a fundamental device of the LSI (Large Scale Integration). As the LSI becomes highly integrated with the needs for high performance, gate oxide film used in the MOSFET is demanded to be as thin as several nano meters. Consequently a small defect of the gate oxide or ununiformity of the gate oxide thickness could affect the electric device performance. Therefore controlling the oxide film in nano scale is essential. In order to manufacture the long lifetime and highly reliable gate SiO_2 films under electric stress, people have been focusing on the understanding of the breakdown mechanism of SiO_2. Though the oxide degradation during electric stressing is discussed with traps that are newly generated, it still has not reached to understanding of the mechanism of oxide breakdown. Our group has been proposing that the microscopic oxide degradation contains two components which we call "Deviating" and

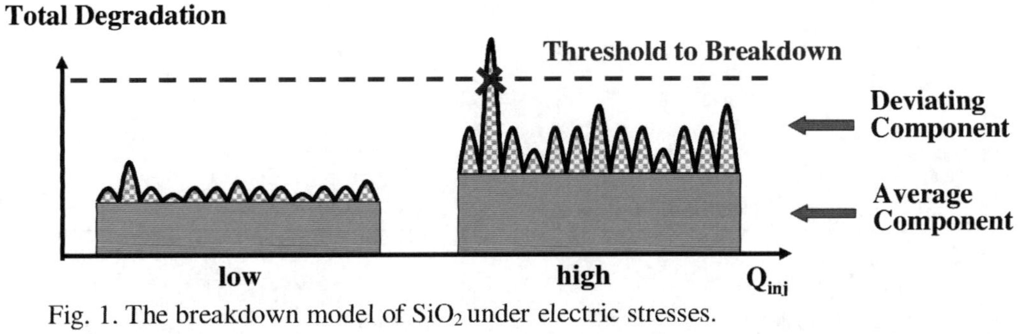

Fig. 1. The breakdown model of SiO_2 under electric stresses.

"Average" components [1], as shown in Fig. 1. These two components grow during the electric stress until the sum of the components reach to the threshold value where the breakdown, i.e. the current runaway occurs. The components have been explained as the internal damage of SiO_2 caused by the high electric field stress, which is distributed two dimensionally. The average degradation can be monitored in a voltage-time (*V-t*) curve under constant current stresses (CCS), for example [2]. In addition, it was proposed that the degradation dose not progress uniformly, which is consistent with the fact that dielectric breakdown of SiO_2 occurs locally. This ununiformity was referred as to deviating component. This component could be attributed to the ununiformity of SiO_2 films, such as surface and interface roughness giving rise to two dimensional thickness fluctuation, and so forth.

Based on the idea of the degradation model, it could be concluded that suppressing both the deviating and the average components is inevitable to achieve longer breakdown lifetime. In this report, to suppress the deviating component, we chose the radical oxidation process, which has been reported to be able to achieve excellent SiO_2/Si interfacial properties compared to the conventional thermal oxides [3-5]. However, the bulk property of radical oxide is still inferior to that of the thermal oxides. Therefore both of the radical oxidation and thermal oxidation processes was used to accomplish the high quality SiO_2 film.

2. Experiments

After cleaning p-type Si(100) wafers, 8 nm-thick SiO_2 films were formed in different three oxidation methods. One is conventional thermal oxidation in wet ambient at 900°C. The sample fabricated in this method will be referred as to ThOx. The second is oxidation using radical source, which will be referred as to Rad. The other is combination of these oxidation processes, that is, radical oxidation followed by thermal oxidation for 30 s. This sample will be referred as to Rad+Thox. Then poly Si gate electrodes with the area of 10^{-4} cm^2 were fabricated on all wafers to form MOS capacitors. The electrical degradation characteristics were investigated by time dependent dielectric breakdown (TDDB) under CCS application. In addition, surface and interface roughness of the SiO_2 films were evaluated by atomic force microscopy (AFM), to relate with degradation characteristics. Further, etching rate of SiO_2 films in HF solution was also characterized.

.

3. Results and Discussion

Figure 2 shows the capacitance-voltage (*C-V*) characteristics of each oxide obtained at 1 MHz. All *C-V* curves were normalized by the saturation capacitance (C_{ox}). Fowler-Nordheim (F-N) plots (acquired from *I-V* measurement) were also indicated in Fig. 3. It is clear that both the *C-V* characteristics and the F-N plots of each oxide were independent of oxidation processes. The flat-band voltage and the conduction band offset of each oxide here is 0.8 5V and 3.5 eV, respectively. The

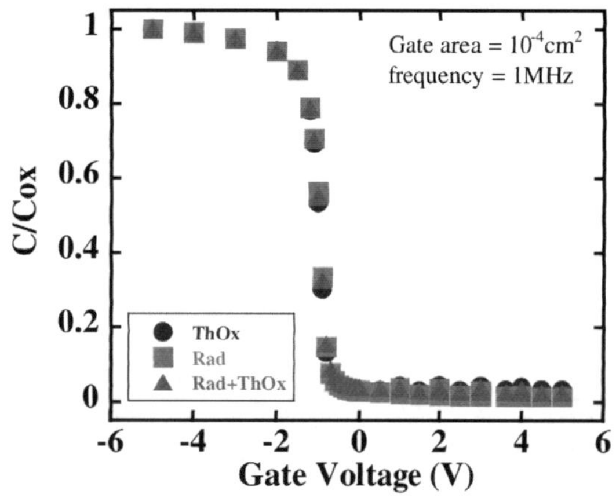

Fig. 2. *C-V* characteristics of each oxide.

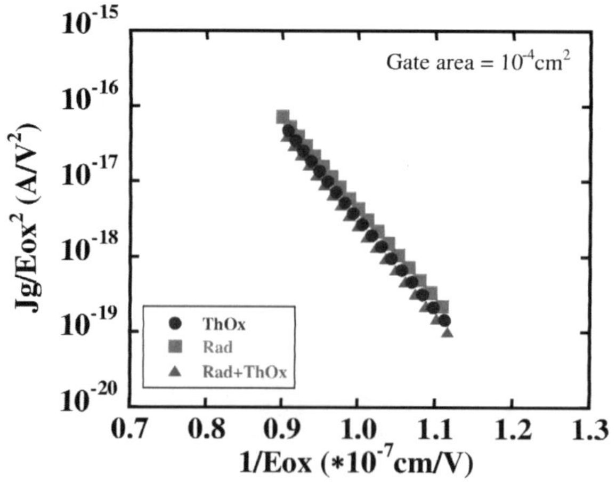

Fig. 3. F-N plots of each oxide.

results indicate that the initial electric characteristics of each oxide are the same.

In order to evaluate the average component, constant current stress was applied. Figure 4 shows the voltage vs. injected charge (*V-Q*) curves of each oxide under constant current stresses of 0.5 A/cm^2 obtained until dielectric breakdown. It is clear that the voltage increases linearly except initial part, and we defined the slope as degradation rate (Deg. Rate) of average component [2]. The Deg. Rate of each curve is indicated in the figure, as 0.0194 and 0.0181 Vcm^2/C for Rad and ThOx, respectively. That is to say Rad has a larger Deg. Rate than that of ThOx. However, it is interesting that the total amount of the gate-voltage shift of Rad is the largest among all oxides, which seems to indicate that the threshold value of degradation for breakdown is highest for Rad. Further, it is also interesting that the curve for Rad+ThOx is quit similar to that for ThOx, indicating that the bulk properties of Rad+ThOx are similar to ThOx's even though additional thermal oxidation is only 30 s. In addition, the Deg. Rate is even smaller than that of ThOx. The results can be attributed to the reconstruction of Si-O network throughout the film by the thermal oxidation process. In other words, thermal oxidation process could have the ability to lead the oxidizer to the whole oxides swiftly. This network reconstruction may be the key point to improve the bulk properties. This will be discussed later.

Figure 5 shows the Weibull plot of the breakdown lifetime of each oxide under constant current stresses of 0.5 A/cm^2. It is clear that the slope of the each Weibull plot is almost same, which means that there is no difference of the breakdown mechanism among the oxides. However it is obvious from the figure that the breakdown lifetime of Rad is longer than that of ThOx even though the Deg. Rate of Rad was larger. The reason can be explained by taking account of the deviating component in degradation.

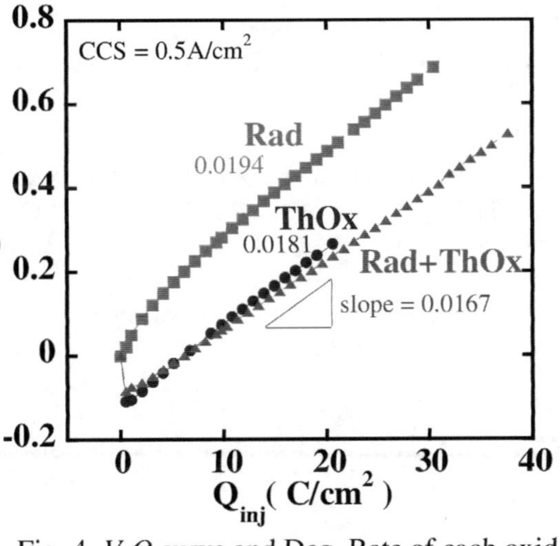

Fig. 4. *V-Q* curve and Deg. Rate of each oxide.

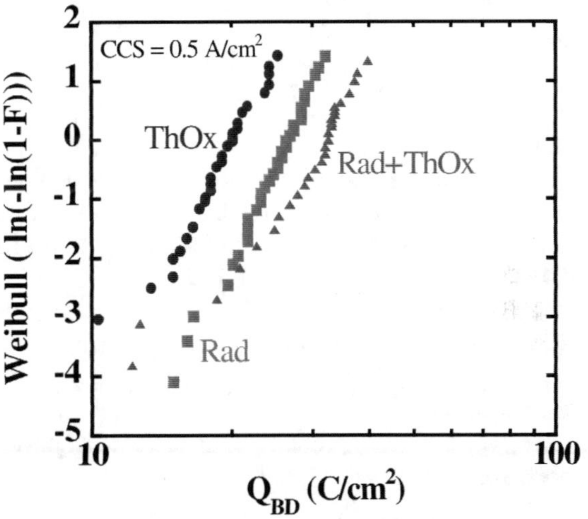

Fig. 5. Weibull plot of the breakdown lifetime.

Figure 6 shows the roughness of the surface (open) and the interface (filled) of each oxide. Here, RMS value represents the degree of the roughness. In other words, the bigger the RMS the poorer the roughness is. For the surface, the RMS value of Rad is slightly smaller than ThOx's. In addition, the value for interface of Rad is about 0.1 nm lower than the ThOx's. The results indicate that uniformity of Rad films, probably thickness uniformity, is superior compared to ThOx. Our conclusion is that this uniformity gives rise to reduction of deviated component in degradation, and it compensates the inferior average component to achieve the longer breakdown lifetime. By the way, the RMS value of Rad+ThOx keeps similar with the Rad's in spite of that the second process is the thermal oxidation. This is probably because the time of the thermal oxidation is so short that the flatness of the SiO_2/Si interface is not ruined.

From these results, it can be concluded that Rad+ThOx inherited both the advantage of the average component of ThOx and the deviating component of Rad, resulting the longest breakdown lifetime as shown in Fig. 5.

Figure 7 shows the etching rate of each oxide in 0.17 % HF solution. It is clear that the etching rate of Rad is larger than that of ThOx throughout the whole oxide thickness. On the assumption that ThOx is fully relaxed, this could be thought that the Si-O network in Rad is more strained compared to ThOx's. The most interesting result in this figure is that the etching rate of Rad+ThOx is similar to ThOx. This indicates that Si-O network in Rad was modulated by short period of thermal oxidation, and the network which is stable against electrical stress was reconstructed.

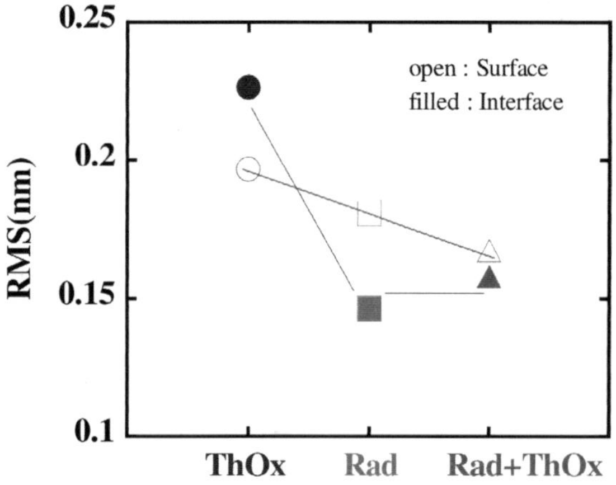

Fig. 6. SiO_2/Si surface and interface

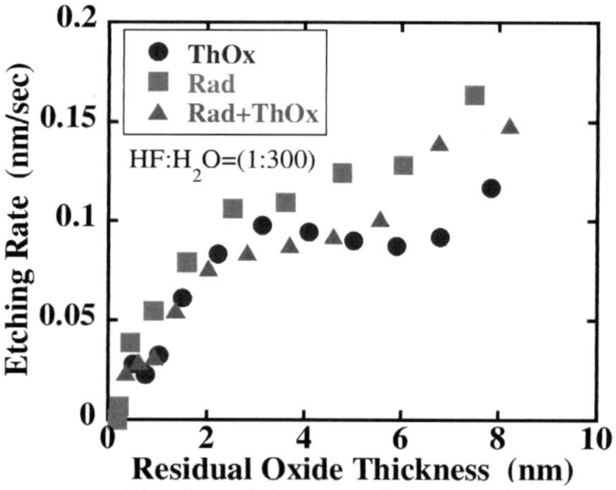

Fig. 7. Etching rate of each oxide.

4. Conclusion

In this research we utilized both the radical and the thermal oxidation to control the deviating and average components individually. The SiO_2 film (Rad+ThOx), which is oxidized firstly by radical oxidation followed by thermal oxidation, would inherit both the advantage of the average component of ThOx and the deviating component of Rad. As a result this two step oxides showed the longer breakdown lifetime compared to the oxides manufactured by the conventional thermal oxidation. This new oxidation method would be a help to improve the quality of SiO_2 films.

Acknowledgement

This work was financially supported in part by a grant from the Satellite Venture Business Laboratories of the University of Tsukuba.

References

[1] K. Yamabe, L. Kai, M. Murata, Jpn. J. Appl. Phys. **38** (1999) L1453

[2] Z. Lu, M. Ogino, H. Kuribayashi, Y. Sugahara, R. Hasunuma, and K. Yamabe, abstract for IWDTF 2008. (2008) 123

[3] Y. Saito, K. Sekine, N. Ueda, M. Hirayama and T. Ohmi, Dig. Tech. Pap., Symp on VLSI Technology, (2000) 176

[4] U. Koji, N. Makoto, I. Hitoshi, T. Akira, Technical report of IEICE. SDM. **100**, No.373 (20001012) 27

[5] K. Sekine, Y. Saito, M. Hirayama and T. Ohmi, IEEE Trans. Elec. Dev., **48** (8) (2001) 1550

AUTHOR INDEX